Digital System
Clocking

DIGITAL SYSTEM CLOCKING

High-Performance and Low-Power Aspects

VOJIN G. OKLOBDZIJA

VLADIMIR M. STOJANOVIC

DEJAN M. MARKOVIC

NIKOLA M. NEDOVIC

The Institute of Electrical and Electronics Engineers, Inc., New York

A JOHN WILEY & SONS PUBLICATION

Published by John Wiley & Sons, Inc., Hoboken, New Jersey.
Published simultaneously in Canada.

For general information on our other products and services please contact our Customer Care Department within the U.S. at 877-762-2974, outside the U.S. at 317-572-3993 or fax 317-572-4002.

Wiley also publishes its books in a variety of electronic formats. Some content that appears in print, however, may not be available in electronic format.

Library of Congress Cataloging-in-Publication Data is available.

ISBN 0-471-27447-X

10 9 8 7 6 5 4 3 2

To Our Parents

CONTENTS

Preface   xiii

Chapter 1   Introduction   **1**

1.1   Clocking in Synchronous Systems   2
1.2   System Clock Design   8
    1.2.1   Global System Clock Generation   9
    1.2.2   On-Chip Clock Generation   11
    1.2.3   Noise Sources and Loop Bandwidth   14
    1.2.4   Design Considerations   15
1.3   Timing Parameters   16
    1.3.1   Clock Skew   16
    1.3.2   Clock Jitter   17
1.4   Clock Signal Distribution   18
    1.4.1   Historical Overview   18
    1.4.2   Clock Distribution in Modern Microprocessors   19

Chapter 2   Theory of Clocked Storage Elements   **27**

2.1   Latch-Based Clocked Storage Elements   27
    2.1.1   True-Single-Phase-Clock Latch   29
    2.1.2   Pulse Register Single Latch   32
2.2   Flip-Flop   34
    2.2.1   Time Window-Based Flip-Flops   41

Chapter 3 Timing and Energy Parameters   **47**

3.1   Timing Parameters   47

     3.1.1   Clock-to-Output Delay, t_{CQ} 47

     3.1.2   Setup Time, U 48

     3.1.3   Hold Time, H 50

     3.1.4   Late Data Arrival and Time Borrowing   52

     3.1.5   Early Data Arrival and Internal Race Immunity   53

     3.1.6   Minimum Data Pulse Width   54

3.2   Energy Parameters   55

     3.2.1   Components of Energy Consumption   55

     3.2.2   Energy Breakdown   57

     3.2.3   Energy per Transition   60

     3.2.4   Glitching Energy   60

3.3   Interface with Clock Network and Combinational Logic   61

     3.3.1   Interface with Clock Network   61

     3.3.2   Interface with Combinational Logic   62

Chapter 4 Pipelining and Timing Analysis   **63**

4.1   Analysis of a System that Uses a Flip-Flop   63

     4.1.1   Late Data Arrival Analysis   63

     4.1.2   Early Data Arrival Analysis   65

4.2   Analysis of a System that Uses a Single Latch   66

     4.2.1   Late Data Arrival Analysis   66

     4.2.2   Early Signal Arrival Analysis   68

4.3   Analysis of a System with a Two-Phase Clock and Two Latches in an M–S Arrangement   70

4.4   Analysis of a System with a Single-Phase Clock and Dual-Edge-Triggered Storage Elements   75

     4.4.1   Late Data Arrival   76

     4.4.2   Early Data Arrival   78

Chapter 5 High-Performance System Issues   **83**

5.1   Absorbing Clock Uncertainties   83

     5.1.1   Clock-Uncertainty Absorption Using Soft Clock Edge   85

     5.1.2   Timing Analysis with Clock-Uncertainty Absorption   88

     5.1.3   Clock-Uncertainty Absorbing Considerations   90

5.2   Time Borrowing   91

    5.2.1   Dynamic Time Borrowing   92

    5.2.2   Static Time Borrowing   96

5.3   Time Borrowing and Clock Uncertainty   97

    5.3.1   Level-Sensitive Clocking   98

    5.3.2   Soft-Edge-Sensitive Clocking   102

Chapter 6   Low-Energy System Issues   105

6.1   Low-Swing Circuit Techniques   108

    6.1.1   Conventional CSEs with Reduced-Swing Clock
        Drivers   109

    6.1.2   CSE Redesign   110

    6.1.3   N-Only CSEs with Low-Supply-Operated Clock
        Drivers   111

6.2   Clock Gating   112

    6.2.1   Global Clock Gating   112

    6.2.2   Local Clock Gating   113

6.3   Dual-Edge Triggering   115

    6.3.1   Latch-Mux Design   116

    6.3.2   Pulsed-Latch Design   117

    6.3.3   Flip-Flop   118

    6.3.4   Clock Distribution   119

6.4   Glitch Robust Design   122

Chapter 7   Simulation Techniques   125

7.1   The Method of Logical Effort   125

    7.1.1   Multistage Logic Networks   126

    7.1.2   Logical Effort of Logic Gates Commonly Found in
        CSEs   127

7.2   Environment Setup   130

    7.2.1   HLFF Sizing Example   134

    7.2.2   M-SAFF Sizing Example   136

    7.2.3   Energy Measurements   137

    7.2.4   Automating the Simulations   138

7.3   Appendix   139

    7.3.1   The CSE Characterization Script   139

    7.3.2   Simulation Bench for FO4 Inverter Delay Extraction
        (simInv.hsp)   146

    7.3.3   CSE Simulation Bench in SPICE (sim.hsp)   148

7.3.4	Example HLFF Deck (hllf16.hsp)	151
7.3.5	Example M-SAFF Deck (saff16.hsp)	153

Chapter 8   State-of-the-Art Clocked Storage Elements in CMOS Technology   **155**

8.1	Master–Slave Latch Examples	155
	8.1.1 Derivation of Master–Slave Latch	155
	8.1.2 C^{2}MOS Master–Slave Latch	158
	8.1.3 Comparison	158
8.2	Flip-Flop Examples	159
	8.2.1 Hybrid-Latch Flip-Flop	159
	8.2.2 Semidynamic Flip-Flop	160
	8.2.3 Sense-Amplifier-Based Flip-Flop	161
	8.2.4 Modified Sense-Amplifier-Based Flip-Flop	163
	8.2.5 Comparison	164
8.3	Clocked Storage Elements with Local Clock Gating	167
	8.3.1 Master–Slave Latch with Local Clock Gating	168
	8.3.2 Data-Transition Look-Ahead Latch	169
	8.3.3 Clock-on-Demand Pulsed Latch	172
	8.3.4 Conditional Capture Flip-Flop	174
	8.3.5 Comparison	176
8.4	Low-Swing Clock Storage Elements	177
	8.4.1 CSE Examples	177
	8.4.2 Comparison	178
8.5	Dual-Edge-Triggered Clocked Storage Elements	180
	8.5.1 DET Latch-Mux	180
	8.5.2 DET C^{2}MOS Latch-Mux	181
	8.5.3 DET Pulsed-Latch	182
	8.5.4 DET Symmetric Pulse Generator Flip-Flop	183
	8.5.5 Comparison	184
8.6	Summary	187

Chapter 9   Microprocessor Examples   **189**

9.1	Clocking for Intel Microprocessors	190
	9.1.1 IA-32 Pentium Pro	191
	9.1.2 First IA-64 Microprocessor	193
	9.1.3 Pentium 4	196
9.2	Sun Microsystems Ultrasparc-III Clocking	200
	9.2.1 Clocking	201
	9.2.2 Storage Elements	202

9.3   Alpha Clocking: A Historical Overview   207
   9.3.1   Clocking   208
   9.3.2   Clocked Storage Elements   212
9.4   Clocked Storage Elements in IBM Processors   217
   9.4.1   Level-Sensitive Scan Design   218
   9.4.2   Examples of Clocked Storage Elements   221

References   **233**

Index   **241**

PREFACE

Is it possible to write an entire book on the subject of clocked storage elements: latches and flip-flops? We certainly did not think so and we are sure many people today share this view. Indeed, this work started as a simple consulting project for Hitachi America Laboratories in the late 1990s that was not intended to last longer than six months. The objective was to examine several proposed and existing clocked storage elements and decide which one should be used in the new generation of microprocessors Hitachi had on the drawing board at that time. We finished this work, comparing several existing structures and recommending some improved solutions. However, the answers we provided raised many more questions and left us wondering. Now we feel that there are even more unanswered questions. Thus, we decided to collect our experience into a book and make it available to design engineers, practitioners, academics, managers, and anyone else interested in this aspect of high-performance and low-power digital system design.

Clocking is an important aspect and a centerpiece of digital system design. Not only does it have the highest positive impact on performance and power, but also the highest negative impact on the reliability of an improperly designed system. This is becoming more important, as the clock frequency keeps increasing dramatically as it has been in the last decade. The higher the frequency, the more important are the clock system and clock storage elements, because their effects do not scale proportionally with other features that are benefiting from the rapid technological advances of the past fifty years. In this book we treat synchronous systems, which we assume will continue to progress in this direction. In reality, we do not know how long this progress will continue. Other ways of timing digital systems are possible, but they have not demonstrated sufficient progress to become a mainstream solution. We do not pretend to know what the timing of digital systems will be in the future, but we hope to provide sufficient analysis and possibly set the stage for the new approaches that will evolve.

This book is divided into nine chapters. In Chapter 1 we provide an overview of clocking and how the clocked storage elements fit into the whole picture. The presentation tends to be historic, as we wish to put the development of clocking and clocked storage elements into needed perspective. Some basic definitions are provided and we tie the clock storage elements into the entire digital system, most particularly into clock generation, distribution testability, and control. Chapter 2 describes clocked storage elements and provides definitions and a clear classification of basic clocked storage elements used in digital systems today. It shows the systematic derivation of flip-flops and sets the stage for the discussion of advanced structures and their performance and energy aspects. The Chapter 3 introduces the timing and energy parameters of the clocked storage elements. Since the speed required for the operation can always be traded for less energy (and vice versa), it is important to tie the two together and place the analysis of performance and power in perspective. Also defined in this chapter is when the data should arrive so that the system operates reliably, as well as the various parameters which affect the power consumption of the system, such as switching activity, voltage scaling, and design style. Chapter 4 provides a rigorous quantitative analysis of clocking. The choice of the clocked storage elements requires a particular analysis of its effects, and the chapter provides various performance and design trade-offs. The quantitative analysis and derivation of the timing parameters for optimal system performance are also presented, starting with the simple flip-flop-based systems and ending with the complex dual clock-edge clocked systems. This chapter should provide the reader with the mathematical tools for determining the optimal system parameters for the design. In order to make these points clear, the chapter ends with examples of two advanced clocking techniques: one for high-performance, and other oriented toward the low-power system. Chapter 5 is dedicated to the issues encountered in designing high-performance systems. Due to the increased effect of clock uncertainties, dealing with the clock skew and jitter and the ability to absorb those unavoidable effects is one of the most important issues in high-performance system design. Since the time boundaries between the stages are more difficult to control precisely, the data from one pipeline stage may take some amount of time from the following one. This subject, also known as *time borrowing* is analyzed, and its relation to clock uncertainty absorption is shown. Chapter 6 is dedicated to low-power system design. It treats the energy issues, in particular, energy reduction. Various ways of achieving low energy per operation, such as supply voltage scaling, reduced signal swing clocking, clock gating, and capturing the data on each transition of the clock signal — *dual-edge triggering* — are described in this chapter. Clocked storage elements designed with features that minimize energy consumption, such as conditional clocking and conditional precharging, are described and analyzed. Chapter 7 describes simulation techniques and optimization methods used to properly size the transistors. It discusses the use of the logical effort technique, and it shows how it is applied to the problem of optimizing clocked storage elements. Most importantly, in this chapter we describe the evaluation setup that should be used in providing a fair comparison between different clocked storage elements and all the miscellaneous issues that affect this comparison. We provide a script used to simulate clocked storage elements in the Appendix to Chapter 7. This script should serve as a starting point for an engineer who is embarking on this elaborate and tedious undertaking, and we hope it will be useful. In Chapter 8 we compare the various clocked storage elements that are commonly known or used in systems with outstanding features, such as high performance or low power. This chapter should provide the reader with a feel for the current state of the art in clocked storage elements and present the designer with possible choices for his or her designs. Finally Chapter 9 describes clocking techniques and clocked storage elements used in representative and well-known microprocessors. It also illuminates various techniques used by microprocessor designers, as well as various design styles and approaches used by different companies that may not be widely known. This chapter summarizes all the knowledge presented in this book and shows the reader how this knowledge is applied by various practitioners in this highly competitive field.

We hope this book will help in achieving even higher microprocessor performance than that available today and set the stage for a number of successful future designs.

VOJIN G. OKLOBDZIJA
VLADIMIR M. STOJANOVIC
DEJAN M. MARKOVIC
NIKOLA M. NEDOVIC

Berkeley, California
October 2002

Digital System Clocking

CHAPTER 1

INTRODUCTION

Clocking is one of the single most important decisions facing the designer of a digital system. Unfortunately much too often it has been taken lightly at the beginning of a design and that viewpoint has proven to be very costly in the long run (Wagner 1988). Thus, it is not pretentious to dedicate an entire book to this subject. However, this book is limited to the even narrower issue of clocked storage elements (CSE), widely known as flip-flops and latches. The issues dealing with clock generation, frequency stability and control, and clock distribution are too numerous to be discussed in depth in this book and so they are covered only briefly. The interested reader is referred to the other books dealing with those issues, such as the one by Friedman (1995).

The importance of clocking has become even more emphasized, as the clock speed is rising rapidly, doubling every three years, as seen in Fig. 1.1. However, the clock uncertainties have not been scaling proportionally with the frequency increase, and an increasingly large portion of the clock cycle has been spent on the clocking overhead. The ability to absorb clock skew or to make the clocked storage element faster is reflected directly in the enhanced performance, since the performance is directly proportional to the clock frequency of a given system. Such performance improvements are very difficult to obtain using traditional techniques on the architecture or microarchitecture levels. The difficulties are caused by the overhead imposed by the CSE delay, and the clock uncertainties. Thus, setting the clock to the right frequency, and utilizing every available picosecond of the critical path, is increasingly important. It is our opinion that traditional clocking techniques will reach their limit when the clock frequency reaches the 5 to 10 GHz range. Thus, new ideas and new ways of designing digital systems are needed. We do not pretend to know what the future trend in clocking should

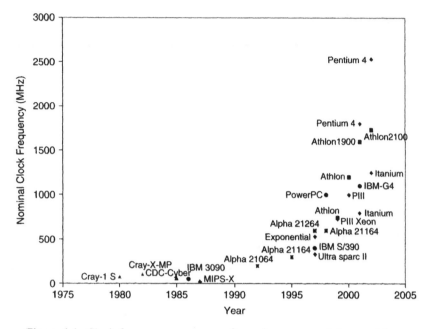

Figure 1.1. Clock frequency versus year for various representative machines.

be, but we feel that some of the ideas discussed in this book can provide a good path to follow.

Computers built in the past were large and filled several electronic cabinets in large air-conditioned rooms that occupied entire floors. They were built from discrete components or used a few large-scale integration (LSI) chips in the later models. Those systems were clocked at frequencies of about one or a few tens of megahertz, as shown in Table 1.1. The first electronic computer, ENIAC (Electronic Numerical Integrator and Calculator), for example, operated at the maximal clock frequency of 18 kHz. Given the low scale of integration, it was possible to "tune" the clock. This was achieved by either adjusting the length of the wires that distributed the clock signals, or by tuning the various delay elements on the cabinets or the circuit boards, so that the clock signal arrived at every circuit board at approximately the same time. With the advent of very large-scale integration (VLSI) technology, and increased integration levels, the ability to tune the clock has been greatly diminished. The clock signals are generated and distributed internally within the VLSI chip. Therefore, much of the burden of absorbing clock signal variations at various points on the VLSI chip has fallen on the clocked storage element.

1.1. CLOCKING IN SYNCHRONOUS SYSTEMS

The notion of clock and clocking is essential for the concept of synchronous design of digital systems. The synchronous system assumes the presence of the

Table 1.1 Clock Frequency of Selected Historic Computers and Supercomputers

System	Date Introduced	Technology	Class	Nominal Clock Period (ns)	Nominal Clock Frequency (MHz)
Cray-X-MP	1982	MSI ECL	Vector processor	9.5	105.3
Cray-1S,-1M	1980	MSI ECL	Vector processor	12.5	80.0
CDC Cyber 180/990	1985	ECL	Mainframe	16.0	62.5
IBM 3090	1986	ECL	Mainframe	18.5	54.1
Amdahl 58	1982	LSI ECL	Mainframe	23.0	43.5
IBM 308X	1981	LSI TTL	Mainframe	24.5, 26.0	40.8, 38.5
Univac 1100/90	1984	LSI ECL	Mainframe	30.0	33.3
MIPS-X	1987	VLSI CMOS	Microprocessor	50.0	20.0
HP-900	1982	VLSI CMOS	Micromainframe	55.6	18.0
Motorola 68020	1985	VLSI CMOS	Microprocessor	60.0	16.7
Bellmac-32A	1982	VLSI CMOS	Microprocessor	125.0	8.0

Source: Wagner 1988.

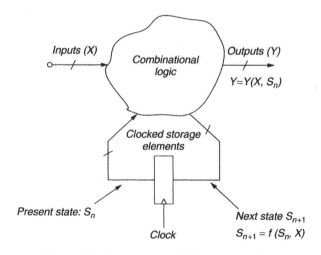

Figure 1.2. The concept of finite-state machine.

storage elements and combinational logic, which together make up a finite-state machine (FSM). The changes in the FSM are in general the result of two events: clock and input signal changes, as illustrated in Fig. 1.2.

The next state, S_{n+1}, is a function of the present state, S_n, and the logic value of the input signals: $S_{n+1} = S_{n+1}(S_n, X_n)$. The remaining question is: When in time will FSM change to the next state, S_{n+1}. This change is determined by the type of clocked storage elements used and the clock signal. The function of the clock signal is to provide a reference point in time when the FSM changes from the present, S_n, to the next state, S_{n+1}. This process is illustrated in Fig. 1.3.

In Fig. 1.3, we have implicitly assumed that the moment when the state changes from S_n to S_{n+1} is determined by the change in the clock signal from logic "0" to logic "1." In fact, this change is determined by the type of clocked storage element and its functionality. We will be discussing this point in detail later in this book. For the purposes of this discussion, we observe that without the clock signal, the change from S_n to S_{n+1} could not be precisely determined. There are digital systems where this change is not caused by the presence, or more precisely, by a change in the clock signal, but by a change of the data signal, for example. Such systems are known as *asynchronous systems*, because they do not require the presence of the clock signal in order to effect an orderly transition from S_n to S_{n+1}. A great deal of research in defining a workable asynchronous system has been done in the last several decades. Recently a microprocessor was designed to operate in an asynchronous manner, and it has been claimed that some small advantages in power consumption were obtained (Woods et al. 1997). In spite of that, the practicality and advantage of the asynchronous design has yet to be proven (Furber et al. 2001). In this book, we limit our discussion to synchronous systems.

If we extend the FSM state diagram in time, we obtain an illustration of the pipeline design (Fig. 1.3). In many cases, when dealing with the synchronous design, the delay throughout the logic block is excessive and the signal change cannot propagate to the inputs of the clocked storage elements in time to effect the change to the next state. In that case, the machine has not met the "critical-path requirement." Such an FSM will fail in its functionality, because the changes

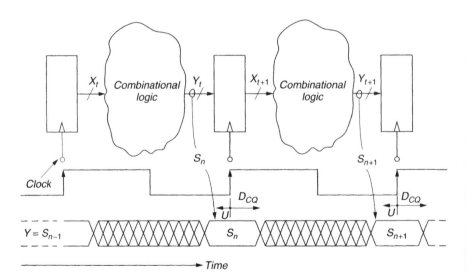

Figure 1.3. State changes in the finite-state machine.

initiated by the input signals will have no effect. This is because the time allowed to change to the next state, S_{n+1}, is too short and the input signal change does not have sufficient time to propagate. In technical jargon this is known as *critical-path violation*. Critical path is defined as the chain of gates in the *longest* (slowest) path through the logic, which causes a signal to take a certain length of time to propagate from the input to the output. Often times, an additional state (or states) is inserted to assure that every transition proceeds in an orderly and timely fashion. This is known as pipelining. A diagram of a pipelined system is shown in Fig. 1.4.

Several clock cycles may be needed in order for the signal to propagate through various stages of a computer system. In general, execution of an instruction may require several *machine cycles*, where machine cycle is defined as the time interval necessary for one atomic operation to execute an instruction. One machine cycle normally takes several clock cycles. The machine cycle is often designated by a waveform defining its own cycle. This is especially true if *microcode* is used to control the machine. In the past, microcoding was a popular concept and it was used extensively in Complex Instruction Set Computers (CISC). In those cases, a process of executing an instruction required several machine cycles. During each machine cycle one *microinstruction* was executed. It normally took several microinstructions to execute an instruction. Each machine cycle required one or several register transfers or passes through several pipeline stages. That in turn required one or more clock cycles, or multiple phases of the clock. Thus, the clocking was quite complex and encompassed several levels of hierarchy. This

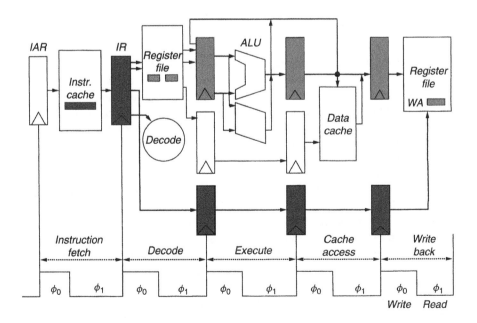

Figure 1.4. Diagram of a pipelined system.

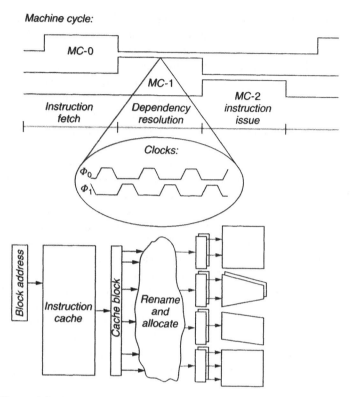

Figure 1.5. Machine execution phases with respect to the clock cycles.

is illustrated in Fig. 1.5, where three distinct machine cycles, *Instruction Fetch*, *Dependency Resolution*, and *Instruction Issue*, are shown. Dependency resolution can be quite a complex operation, requiring several *register transfers*, which means several clock cycles are necessary to complete this operation (as shown in Fig. 1.5). The machine would normally scan the cache block for several instructions and try to resolve any data dependencies. At the end of this cycle, operands will be fetched and placed in the corresponding registers (reservation stations) of the execution units.

In microcoded machines a large disparity existed between the speed of the clock and the speed of logic. It could take several clock cycles or even several tens or hundreds of clock cycles to execute one instruction. A more complex instruction required many more clock cycles. There could be tens of *logic levels* in the critical path, and 40 to 50 were not uncommon. Thus, the time associated with the clock and clocking was not as critical as it is today.

As the level of integration increased, combined with the increased speed of today's machines, the number of logic levels in the critical path started to diminish rapidly. Today's high-speed processors are either implementing Reduced Instruction Set Computer (RISC) architecture, or are running CISC code. However, to be able to efficiently implement superscalar execution cores, even CISC computers are translating their instructions into simple RISC-type operations called ROPs (RISC operations). Their microarchitecture can execute one or several ROPs in place of one CISC instruction. Therefore, the concept of microcoding has disappeared, as did the concept of machine cycle when implementing a particular machine architecture. The instructions (or ROPs) are executed in one cycle, which is usually driven by a single-phase clock. In other words, one instruction (or one ROP) is executed in every clock cycle. The levels of hierarchy that existed between the clock cycle and instruction execution no longer exist. In addition, the number and depth of pipeline stages keeps increasing in order to accommodate the trend toward increasing speed. As a result, the number of logic stages between the two CSEs keeps decreasing. Today 10 levels of logic in the critical path are more common. This number is still decreasing, as illustrated in Fig. 1.6. Any overhead associated with the clock system and clocking mechanism directly and adversely affects machine performance and is therefore critically important.

With this introduction we should be able to understand the function of the clock signal before we proceed with other definitions. The function of the clock signal is comparable to the function of the metronome in music. Similarly, in the digital system the clock designates the exact moment when the state is changing, as well as when the next state is to be captured. Also, all of the logic operations have to finish before the tick of the clock, because their final values are being captured by that clock event. Therefore, the clock provides the time reference point, which determines the flow of the data in the digital system.

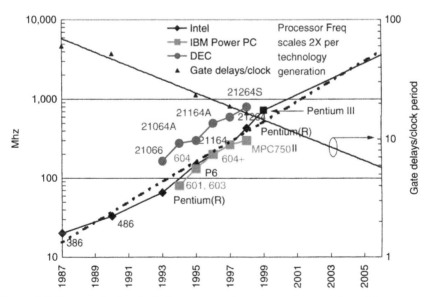

Figure 1.6. Increase in the clock frequency and decrease in the number of logic levels in the pipeline. (Borkar 1999), Copyright © 1999 IEEE.

1.2. SYSTEM CLOCK DESIGN

The clock system is usually divided into two distinct categories: *clock genera-tion* and *clock distribution*. However, this classification should be extended by adding CSEs as an additional category, because the nature of the clocked storage elements is intimately connected to the clock system generation and distribution, and it is the nature of clocked storage elements that dictates the requirements imposed on the clock system. This relationship is best illustrated by the choice of clocking scheme, as shown in Fig. 1.7. The clock system can consist of a single-phase, a two-phase, or a multiple-phase clock. Transfer of data between CSEs in the system is usually accomplished by using an active phase of the clock. Thus, the clock phase controls the transfer of the information among the CSEs in the system. To prevent data from moving further then desired (achieving *nontransparency*), the clock phases are separated in time. This is referred to as *nonoverlapped* clock phases. In high-performance systems various phases of the clock can be overlapped in order to increase total system performance.

In the older systems it was more common to use multiple-phase clocks (Siewiorek et al. 1982). Transparent latches or flip-flops triggered by short pulses were used as storage elements. As the frequency of the operation kept increasing,

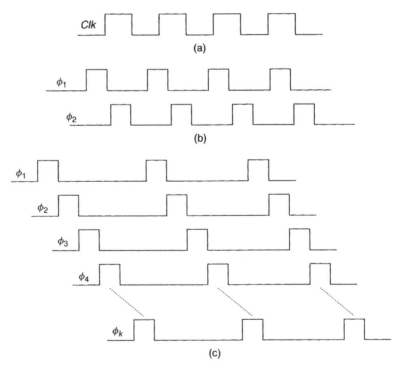

Figure 1.7. System clocking schemes: (a) single-phase clock; (b) two-phase clock; (c) multiple-phase clock.

it became exceedingly difficult to control various phases of the clock and their relationship to each other.

The two-phase clock is a robust scheme and is compatible with the design for testability, a desired feature of a complex computer system. Such a scheme, which incorporates a test mode, has been used in generations of IBM mainframe computers as a part of level-sensitive scan design (LSSD) methodology (LSSD 1985). The two nonoverlapping phases of the clock assure a robust clocking system that can tolerate manufacturing and process-parameter changes.

Given the continuing search for more speed and increased level of integration, even the two phases of the clock became difficult to control on the VLSI chip. This led to the widespread adoption of the single-phase clock in use today. Although two-phase clocking is still used, it is a single-phase clock that is distributed throughout the system, allowing the two necessary phases to be generated locally. This technique achieves two goals: (1) necessary amplification of the clock signals and ability to drive a large row of storage elements (register, for example), and (2) generation of two clock phases and compatibility with scan methodology. A scheme used for local two-phase clock generation from a single-phase clock distributed on the chip is shown in Fig. 1.8. Such a scheme is also capable of supporting the test and debug mode. The two phases of the clock, C_1 and C_2, are generated from the global clock CLKG. Specialized circuitry was added to allow for edge shifting at the cycle boundary (Sigal et al. 1997). Enabling and disabling of the clock phases is used to switch from normal operation to the *scan* mode that is used for testing.

1.2.1. Global System Clock Generation

Clock generation begins on a system board, where the global system clock reference is generated from a "crystal" oscillator. This is a circuit that uses

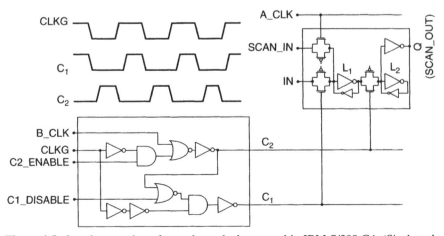

Figure 1.8. Local generation of two-phase clocks as used in IBM S/390 G4. (Sigal et al. 1997), reproduced by permission.

a piezoelectric quartz crystal or some other ceramic material as a mechanical representation of an electrical inductance–capacitance–resistance (*LRC*) series resonant circuit. Piezoelectric effect in a material occurs with the exchange of energy between the mechanical forces and applied electric field. In quartz crystal, the physical dimensions of the lattice can very precisely determine the oscillation frequency. One excellent property of such resonators is their extremely high Q-factor, typically 1000–10,000. By attaching a nonlinear element (such as an NFET) to the resonator, the series resistance of the resonator is canceled by the negative resistance of the nonlinear element and "lossless" oscillations are maintained. Due to the high-quality Q-factor, the variation of the resonant frequency of the oscillator is only a few parts per million (ppm). Two realizations of the clock oscillator are shown in Fig. 1.9a and 1.9b.

System clock is set to directly correspond to the speed of data busses on the system board, that is, from 66 MHz, 100 MHz, 133 MHz, 266 MHz, and higher, in PC boards, to a few hundred MHz in specialized systems. However, the on-chip clocks operate at frequencies that are in the GHz range. Even if the on-board clock signal of the same frequency as the on-chip clock could be generated, it would be very hard to bring it on-chip because of large parasitic capacitances and inductances in the package and bond-wires/balls that connect to the die. For these reasons, the low-frequency system clock is first brought on-chip and then frequency multiplication is performed to achieve the desired on-chip clock rate.

The time difference between the external clock and the internal clock, called *insertion delay* (shown in Fig. 1.10), increases relative to the clock period with the increase in the clock frequency. Input data are synchronized with the external clock, but can be stored directly in the storage elements clocked by the internal clock. Any insertion delay between the external and internal clocks directly

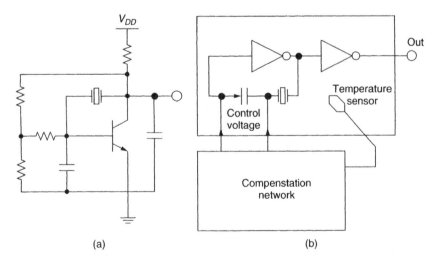

(a)   (b)

Figure 1.9. (a) Crystal oscillator. (b) Temperature-compensated crystal oscillator.

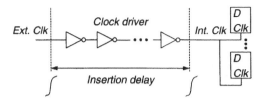

Figure 1.10. On-chip clock insertion delay.

impacts the cycle time of the processor. The insertion delay is caused by the on-chip clock-driver delay, with the inverter chain representing the equivalent of the clock-driver tree, and clocked storage elements representing the total clock load. Several nF of the clock load are routinely encountered in modern microprocessor designs (Young et al. 1992). The clock-driver tree requires five or more fan-out of 4 (FO4) delays, which easily accounts for over 50% of the processor cycle time. Moreover, due to process and environmental variations, the delay of the clock driver may vary, causing an unknown phase relationship of the external and internal clocks.

The problem of external and internal clock alignment can be solved by using the phase-locked loop (PLL). The main task of the PLL is to align the external reference clock with the on-chip internal clock at the end of the clock driver, thus effectively removing the driver delay.

1.2.2. On-Chip Clock Generation

There are two main types of PLLs. In the first type, the PLL has its own voltage-controlled oscillator (VCO) that generates the internal clock, which is then aligned to the external reference clock by the virtue of negative feedback, as shown in Fig. 1.11. The phase difference between the external reference clock and the internal distributed clock is detected with the phase detector (PD), and low-pass filtered (LP), to create the control voltage for the VCO, steering the oscillation frequency in order to align the external and internal clocks, ideally achieving a zero phase difference. At this point, a so-called *phase lock* is achieved (Gardner 1979). This type of PLL was introduced first, and so historically it kept the name PLL. One example of the PLL operation is shown in Fig. 1.11, where the output of the phase detector is the XOR of the external clock reference and the internal clock, producing pulses, p, that are then low-pass filtered to produce the slowly changing control voltage, cv, which changes the frequency of the VCO, and hence the internal clock. At first, the external and internal clocks have a phase difference of 135°, but after the phase difference is detected and the frequency of the internal clock changes, the phase difference is decreased to 45°.

The other type of PLL is delay-line based or delay-locked loop (DLL). As shown in Fig. 1.12, the VCO in the PLL is replaced by the voltage-controlled delay line (VCDL), which delays the external clock, feeding the clock driver,

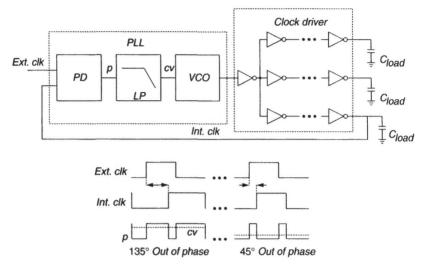

Figure 1.11. Phase-locked loop block diagram and operation.

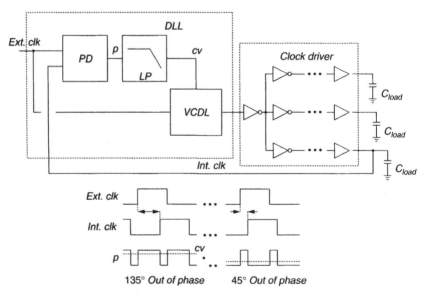

Figure 1.12. Delay-locked loop block diagram and operation.

until the internal clock becomes aligned with the external clock, at which point the control voltage of the VCDL become steady and the loop stays in lock. An example similar to that in Fig. 1.11 is given in Fig. 1.12. The main difference between the examples is that, unlike in Fig. 1.11, the internal clock in Fig. 1.12 does not change frequency over time, but is delayed in order to achieve phase alignment. The key point to understand is that alignment is possible in both PLL and DLL, because both the external and internal clocks are periodic, which delays them by an integer number of cycles with respect to each other, resulting in cancellation of the phase difference. Otherwise, it would not be physically possible to eliminate this delay. It is only possible to add more delay until the total delay becomes an integer number of clock cycles.

In addition to clock alignment, PLLs can perform frequency multiplication. Figure 1.13 shows a general block diagram where the VCO operates at $f_{vco} = f_{ext} \times B \times C/A$, and the frequency of the internal clock is $f_{int} = f_{vco}/B$. Typically, the value of B is two, to guarantee a 50% duty cycle of the internal clock, and the value of A is one. The value of C is set to the ratio between the desired internal-clock frequency and the external (system) -clock frequency (Young et al. 1992), which is always conveniently set to be an integer value, preferably base two. There are, however, cases where multiple values of A, B, and C are used in the power-up sequence to avoid excessive supply noise on large chips, like Alpha 21264 (von Kaenel et al. 1998).

From the standpoint of noise performance, the VCO (VCDL) is the most critical part of the PLL (DLL). It is therefore illustrative to compare most common design styles and discuss the possible trade-offs. VCO is built either as a ring oscillator topology, Fig. 1.14, or an inductance–capacitance (LC) tank oscillator, Fig. 1.15. Ring-oscillator-based VCOs are relatively easy to implement, and require much less area than LC tank oscillators. By regulating the supply,

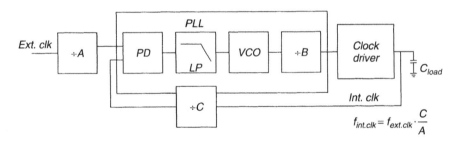

Figure 1.13. PLL frequency multiplication.

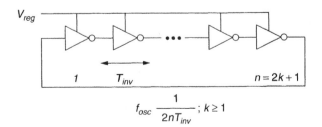

Figure 1.14. Ring-oscillator-based VCO, with CMOS inverters as delay elements.

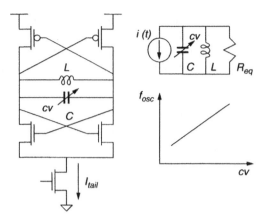

Figure 1.15. *LC* tank-based VCO, equivalent ac circuit model and current waveform.

inverter delay is controlled, and so is the oscillation frequency. The minimum number of stages needed to sustain oscillations is three, since it provides sufficient delay, while typical numbers range from three to seven or more stages, (Hajimiri 1998).

With the increase in clock frequency and the use of on-chip spiral inductors, both feasible with today's technology, *LC* tank-based VCOs are becoming increasingly popular due to superior phase-noise performance. However, *LC* tank oscillators do not always perform better than ring oscillators. This largely depends on the dominant source of noise and the number of stages in the output buffer and ring oscillator (Hajimiri 1998). A typical *LC* tank VCO is shown in Fig. 1.15, with an equivalent small-signal model and frequency characteristic as a function of applied *cv*.

A VCDL can be built from the same delay elements as the ring-oscillator VCO. The delay elements most often used are differential pairs, which provide good power-supply rejection, and the recently popular inverters with a power-supply regulator that performs power-supply filtering and effectively shields the inverters from any power-supply noise (von Kaenel et al. 1998; Sidiropoulos et al. 2000). For details on other PLL and DLL building blocks, see, among others, Gardner (1979), Kim et al. (1994), and Razavi (1996). The following section briefly describes some of the most important noise sources and trade-offs involved in PLL and DLL design, and gives a comparative analysis of PLL versus DLL performance.

1.2.3. Noise Sources and Loop Bandwidth

For the purposes of high-level analysis, we divide the noise sources into three main categories: (1) noise of the reference clock, (2) noise induced in the VCO (or VCDL), and (3) noise induced on the clock during distribution from the PLL (DLL) to the CSE, here defined as *clock driver noise*. Since these noise sources are introduced into the loop at different locations, the transfer functions to the output are different for each of them. For example, input reference noise is low-pass filtered at the output of the PLL, with the filter bandwidth set by the bandwidth of the PLL. On the other hand, input reference noise passes directly through the VCDL to the output of the DLL, without any filtering. Noise induced in the VCO is fed back to the VCO input (in ring-oscillator implementation) and "accumulated" over time (Kim et al. 1994). Any noise induced in the VCO or VCDL is tracked and rejected by the loop, up to the loop bandwidth. Therefore, the transfer function of noise from the VCO (VCDL) to the output is high-pass, contrary to the one from the input reference to the output, which is low-pass. This immediately points to the possible trade-off between the amount of input reference noise and VCO noise at the output of the PLL. Indeed, the optimal bandwidth at which these two noise sources are balanced exists and minimum total noise is achieved (Lim et al. 2000; Mansuri and Yang 2002). In summary, DLLs perform better in cases where the reference clock is not the main source of clock uncertainty and most major noise comes from the noise induced in the VCDL line. PLLs are, however, better in cases where the input reference noise is dominant, and typically worse in cases where the major noise is induced in the VCO, due to the noise accumulation effect, given that compared VCOs and VCDLs are implemented using the same type of delay element.

The preceding analysis is somewhat blurred in modern systems, due to the noise induced in the clock driver. While VCOs and VCDLs are typically implemented using three to seven delay stages, because of the increasing amount of clock load, clock driver depth has increased from generation to generation, and is now over five stages in modern processors. Given that sensitivity of the delay elements in VCO or VCDL is typically an order of magnitude better than that of inverter, which has a 1% delay variation for a 1% power supply variation, it can be easily seen that the overall noise of the distributed on-chip clock is usually dominated by the noise induced in the clock driver tree.

1.2.4. Design Considerations

Regarding the design of the PLLs and DLLs, PLLs are typically harder to design, due to stability issues (PLL is a second-order system due to the integrating function of the VCO), but offer more flexibility than DLLs, that is, wider locking range and, frequency multiplication. DLLs are simpler to design, given that they are first-order systems (unconditionally stable), but offer limited lock range. However, it is true that more complicated DLLs that offer similar flexibility to PLLs are also very complex systems (Sidiropoulos and Horowitz 1997).

PLLs are mostly used in modern processors to multiply the frequency of the external system clock and reject any existing high-frequency reference clock noise. DLLs have recently found application as deskewing elements in high-performance processors, synchronizing different clock domains on a die to the global clock reference from the PLL (Rusu and Tam 2000; Xanthopoulos et al. 2001). It should be noted, however, that these approaches only deal with the DC portion of the noise on the clock (skew), while AC portion of the noise (jitter) is not eliminated. The jitter induced in the clock driver by power supply variations still presents the dominant source of noise in the on-chip clock distribution and needs to be budgeted for in any clocking methodology.

1.3. TIMING PARAMETERS

It is appropriate at this point to consider the clock distribution system and define the clock parameters that will be used throughout this text. For the purposes of definition we should start with the Fig. 1.16, which shows the timing parameters for a single-phase clock.

The clock signal is characterized by its *period*, T, which is inversely proportional to the *clock frequency*, f. The time during which the clock is active (assuming logic 1 value) is defined as *clock width*, W. The ratio of W/T is defined as *clock duty cycle* (w). Usually, the clock signal has a symmetric shape, which implies a 50% duty cycle. This is also the best we can expect, especially when distributing a high-frequency clock. Another important point is the ability to precisely control the duty cycle. This point is of special importance when each phase of the clock is used for logic evaluation, or when we trigger the clock storage elements on each edge of the clock (as we will see later in the book). Some recently reported work demonstrates the ability to control the duty cycle to within ±0.5% (Bailey and Benschneider 1998).

There are two other important timing parameters that we need to define: *clock skew* and *clock jitter*.

1.3.1. Clock Skew

Clock skew is defined as a *spatial variation* of the clock signal as distributed through the system. The clock skew is measured from some reference point in the system; the clock entry point to the board or VLSI chip, or the central point from where the clock distribution starts. Because of the various delay characteristics of the clock paths to the various points in the system, as well as different loading

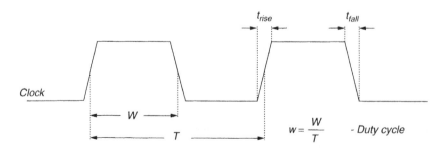

Figure 1.16. Clock parameters: period, width, rise, and fall times.

of the clock signal at different points, the clock signal arrives at different points at different times. This clock skew is defined as the difference between the reference point and the particular destination CSE. Further, we can distinguish *global clock skew* and *local clock skew*. We define global clock skew as the maximal difference between two clock signals reaching any of the two storage elements on the chip, or in the system, that exchange data under the control of the same clock. Our definition of the clock skew describes global clock skew. Clock skew occurring between two adjacent CSEs represents local clock skew. If the two adjacent clock storage elements are connected with no logic in-between, the problem of data race-through can occur. Characterizing a maximum local clock skew is therefore important. These clock skew definitions are equally important in high-performance system design.

1.3.2. Clock Jitter

Clock jitter is defined as *temporal variation* of the clock signal with regard to the reference transition (reference edge) of the clock signal, as illustrated in Fig. 1.17. Clock jitter represents edge-to-edge variation of the clock signal in time. As such, clock jitter can also be classified as *long-term jitter* and *cycle-to-cycle* (or *edge-to-edge*) *jitter*. Edge-to-edge clock jitter is the clock signal

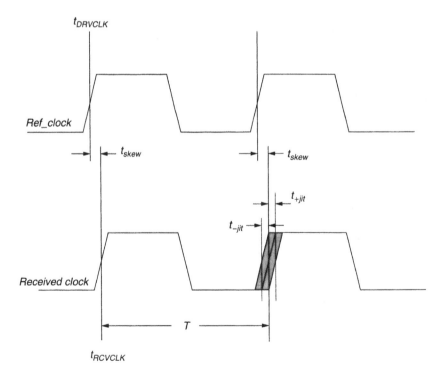

Figure 1.17. Clock parameters: period, width, clock skew, and clock jitter.

variation between two consecutive clock edges. In the course of high-speed logic design, we are more concerned about cycle-to-cycle clock jitter, because it is this phenomena that affects the time available to the logic. Long-term jitter represents clock-edge variation over a large number of clock cycles (long-term). While short-term jitter is dependent on the type and quality of the clock generator, long-term jitter is a result of the accumulated effects. Long-term jitter usually affects communication and synchronization between various blocks within a system that are same distance apart and need to operate in synchrony.

1.4. CLOCK SIGNAL DISTRIBUTION

1.4.1. Historical Overview

Usually a clock signal was generated using a quartz-crystal-controlled oscillator that provides an accurate and stable frequency. Given the size limitation of the quartz crystal, the frequency of such a generated clock signal cannot be very high, and frequencies in excess of 30–50 MHz are rarely generated using a quartz crystal. The clock signal is then conditioned and amplified to reach desirable driving strength before it is applied to the outside pins of a VLSI chip, from which it drives an internal PLL or DLL. Before reaching the boundaries of the VLSI chip, adjustments to its shape and form are possible. In contrast, in older computer systems, which consisted of several electronic cabinets distributed over the computer floor, and which contained a number of printed circuit boards, adjustments to the clock signal were made at each level. Thus, the clock signals were distributed over longer distances and over several levels, including the cabinet, printed circuit boards, and internal modules. Those separate entities entered by the clock signal were referred to as "logic islands," a term introduced by Amdahl (Flynn and Amdahl 1965; Kogge 1981). The concept of logic islands is illustrated in Fig. 1.18.

Figure 1.19 shows that further tuning and delay adjustment of the clock signal is possible at the point where the clock enters the board or cabinet (called an island). Those elements are usually called *tuning points*. The positioning of tuning points in the system is illustrated in Fig. 1.19. Various clock shaping, forming, and tunable delay elements are employed, and some of them are illustrated in Fig. 1.20. These elements make it possible to control the timing of the leading as well as the trailing edge of the clock signal, and to produce an early as well as late clock signal with reference to the nominal clock.

By adjusting the clock delay and subsequently shaping the edges of the clock signal, it is possible to create early, nominal, and late clocks, as shown in Fig. 1.21c. Those clocks then can be routed to various points on the board. Older systems had much greater control of the clock signal than what is possible today, because once the clock reaches the boundary of the LSI chip, tuning and shaping the clock is not possible. This is because it is much more difficult to tune on the chip due to the lack of external control and greater parameter variations

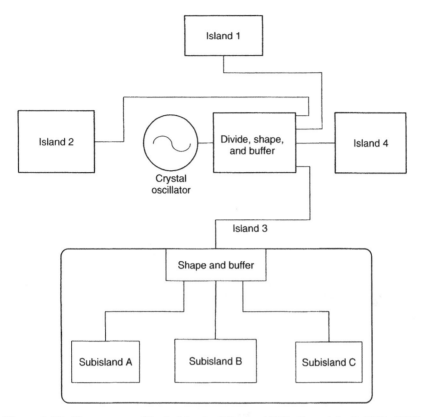

Figure 1.18. The concept of logic islands. (Wagner 1988), Copyright © 1988, IEEE.

on the chip. It is also difficult to build tuning elements such as inductors on the chip and to make adjustments from outside.

With the advent of integration, the systems have shrunk dramatically in size. Today, it is quite common for a processor to have several levels of cache memory contained entirely on a VLSI chip. The chip's capacity for hundreds of millions of transistors makes it possible to integrate not only one processor but also a multiprocessor system onto a single chip. The inability to introduce tuning elements on the chip further aggravates the problem of distributing the clock signals precisely in time, since it is not possible to make further manual adjustment to the clock signal once it has crossed the boundaries of the VLSI chip. Therefore, careful planning and design of the on-chip clock distribution network is one of the most critical tasks in high-performance processor design.

1.4.2. Clock Distribution in Modern Microprocessors

Typically, the clock signal has to be distributed to several hundreds of thousands of the clocked storage elements (flip-flops and latches) on a complex processor

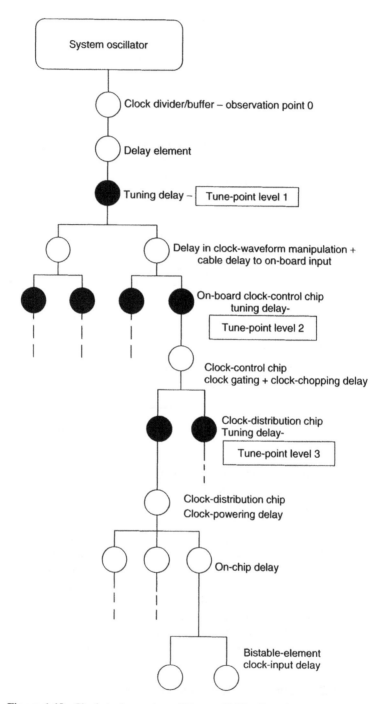

Figure 1.19. Clock tuning points. (Wagner 1988), Copyright © 1988 IEEE.

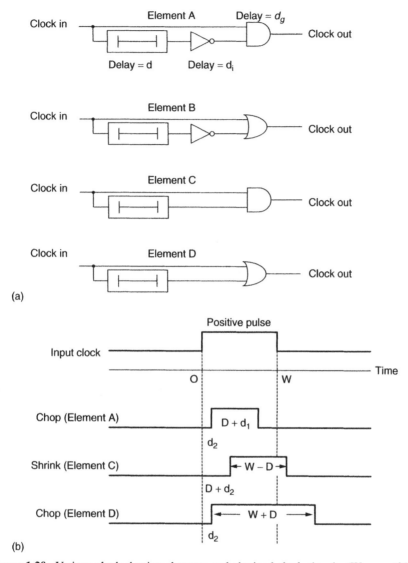

Figure 1.20. Various clock shaping elements and obtained clock signals. (Wagner 1988), Copyright © 1988 IEEE.

chip. Therefore, the clock signal has the largest fan-out of any node in the design, which requires several levels of amplification (buffering). One consequence of imposing such a load on the clock signal is that the clock system by itself can use up to 40–50% of the power of the entire VLSI chip (Gronowsky et al. 1998). However, power is not the only problem associated with the distribution of the clock signals. Since we are dealing with synchronous systems, we must assure that every clocked storage element receives the clock signal at precisely

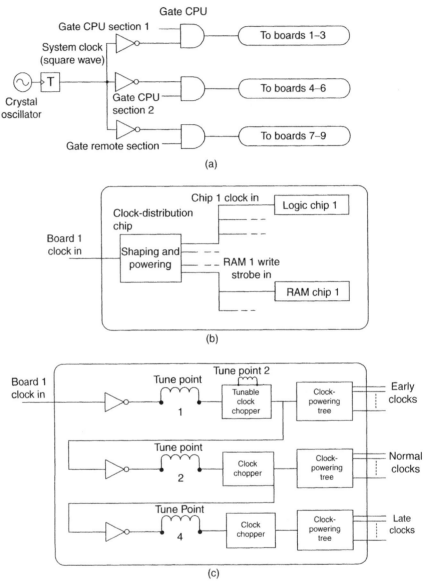

Figure 1.21. (a) Clock distribution network within a system, (b) on the board, and (c) tuning of the clock. (Wagner 1988), Copyright © 1988 IEEE.

the same moment. The clock signal traverses different paths on the VLSI chip, while tracing its path from its origin, the entry point to the VLSI chip, to different clocked storage elements receiving it. Those paths can differ in several attributes, such as the length of the path (wire), the physical properties of the material along different paths, the differences in clock buffers on the chip as a consequence of the process variations. The negative effect of these variations on the synchronous design is that different points on the chip will receive the clock signal at different moments, which results in a further increase in both local and global clock uncertainties.

There are several methods for the on-chip clock signal distribution that attempt to minimize the clock skew and to contain the power dissipated by the clock system. The clock can be distributed in several ways, two of which are worth considering here: (1) resistance–capacitance (*RC*) matched tree shown in Fig. 1.22a, and (2) the grid shown in Fig. 1.22b.

An *RC* matched tree is a method of assuring (to the best of our abilities) that all the paths in the clock distribution tree have the same delay, which includes the same *RC* of the wire, as well as the same number of equal-size buffers on the clock signal path to the storage element. There are several different topologies used to implement an *RC* matched tree. The common objective is to do the best possible in balancing various clock signal paths across the various points on the VLSI chip. An example of four different topologies, as taken from Bailey (Chandrakasan et al. 2001), is shown in Fig. 1.23.

If we had superior computer-aided design (CAD) tools, a perfect and uniform process, and the ability to route wires and balance loads with a high degree of flexibility, a matched *RC* delay clock distribution (Fig. 1.23) would be preferable to grid (b) as shown in Fig. 1.22b and Fig. 1.24. However, none of that is true. Therefore the grid is used when clock distribution on the chip has to be very precisely controlled, which results in higher clock power, as is the case in high-performance systems. This is not difficult to understand given that in a grid arrangement a high-capacitance plate is driven by buffers connected at various points.

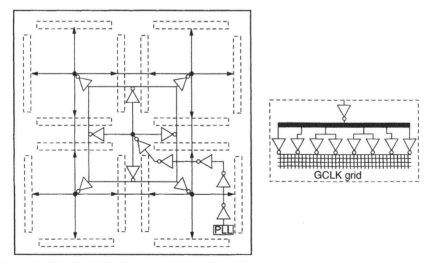

Figure 1.22. Clock distribution methods: (a) an *RC* matched tree, and (b) a grid. (Bailey and Benschneider 1998), Copyright © 1998 IEEE.

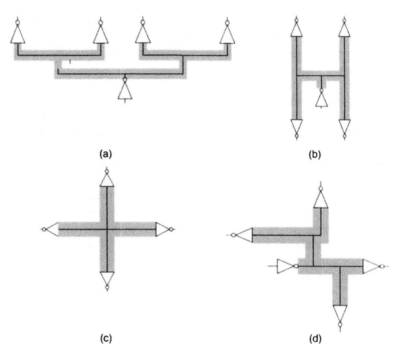

(a) (b)

(c) (d)

Figure 1.23. *RC* delay matched clock distribution topologies: (a) a binary tree (b); an H tree; (c) an X tree; (d) an arbitrary matched *RC* matched tree. (From Bailey in Chandrakasan et al. 2001), Copyright © 2001 IEEE.

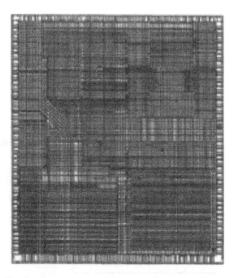

Figure 1.24. Clock distribution grid used in a DEC Alpha 600-MHz processor. (Bailey and Benschneider 1998), Copyright © 1998 IEEE.

One such example is the DEC Alpha processor, which was the fastest processor for several generations of microprocessors starting with the first 200-MHz design introduced in 1992 and ending with the 600-MHz design in 1998 (Dobberpuhl et al. 1992; Benschneider et al. 1995; Gieske et al. 1997). A picture of the clock distribution grid is shown in Fig. 1.24.

With an increased number of transistors, local variations in device geometry and supply voltage become a more important component of the clock uncertainty, which cannot be compensated for by layout (Schutz and Wallace 1998). A more sophisticated clock distribution than simple RC-matched or grid-based schemes is therefore necessary. One such example will be described in Chapter 9 of this book. The active schemes with adaptive digital deskewing typically reduce clock skew of the simple passive clock networks by an order of magnitude, allowing for more tightly controlled clock period and higher clock rates. The digital deskewing circuit for clock distribution evens out the *static components of skew* (load, interconnect, and device mismatches). Additionally, it compensates for the *dynamic* variations in temperature and voltage gradients during all phases of active microprocessor operation.

CHAPTER 2

THEORY OF CLOCKED STORAGE ELEMENTS

The function of a *clocked storage element* is to capture the information at a particular moment in time and preserve it for as long as it is needed by the digital system. Having said this, it is not possible to define a storage element without defining its relationship to a clocking mechanism in a digital system, which is used to determine discrete time events. This definition is general and should include various ways of implementing a digital system. More particularly, the element that determines time in a synchronous system is the *clock*.

2.1. LATCH-BASED CLOCKED STORAGE ELEMENTS

The simplest storage element consists of an inverter followed by another inverter, which provides positive feedback, as shown in Fig. 2.1a. The information bit at the input is thus locked due to the positive feedback loop, and it can be only changed "by force" (i.e., by forcing the output of the feedback inverter to take another logic value). This configuration is used very frequently, and is also known as the *keeper*, a circuit that *keeps* (preserves) the information on a particular node.

If we were to avoid the power dissipation associated with overpowering (forcing) the keeper to change its value, we must introduce nodes that will help us in changing the logic value stored in the feedback loop. For that purpose we are free to use logic NAND or NOR gates, as shown in Fig. 2.1. Of particular interest is a simple modification of the diagram that emphasizes the sum-of-products (SOP) nature of this logic topology. We start with a simple cross-coupled inverter pair, which is unrolled to better illustrate the existing positive feedback (Fig. 2.1a). In the second step we replace the inverters with NAND gates, which enables us to control the variable inside the loop and to selectively set it to 1 or 0 using

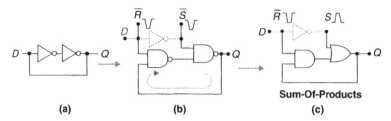

Figure 2.1. Latch structure: (a) keeper; (b) $S-R$ latch; (c) SOP latch.

the input that controls the S and R gates in this case (as shown in Fig. 2.1b). Finally we apply De Morgan rules, which allows us to transform this structure into AND-OR topology. It is well known in digital design that this topology represents SOP, which is a general expression for any Boolean function. The existence of this topology leads to the Earl's Latch (Earl 1965).

It is easy to derive a Boolean equation to represent the behavior of the presented $S-R$ latch. The next output, Q_{n+1}, is a function of the Q_n, S, and R signals. Later in this book we will use those simple dependencies in order to design improved clocked storage elements. The $S-R$ latch can change the output, Q, at any time. In order to make the latch compatible with the synchronous design, we will restrict the time when Q can be affected by introducing the clock signal that gates the S and R inputs. If the data input, D, is connected to S and the property of the $S-R$ latch, which makes S and R mutually exclusive, is applied, the resulting D-latch is shown in Fig. 2.2a. The related timing diagram of a D-latch is shown in Fig. 2.2b. The latch is *transparent* during the period of time the clock is *active*, i.e., assuming logic 1 value.

A latch can be built in a SOP topology (Fig. 2.1c). This tells us that it is possible to incorporate logic into the latch, given that the SOP is one of the basic realizations of the logic function. This leads to the construction of Earl's Latch, which was introduced during the course of the development of a well-known IBM S360/91 machine (Earl 1965; Flynn 1966; Amdahl 1964; Anderson et al. 1967). The basic Earl's Latch configuration is shown in Fig. 2.3a, (Earl 1965), while a latch implementing the Carry function is shown in Fig. 2.3b (Halin and Flynn 1972).

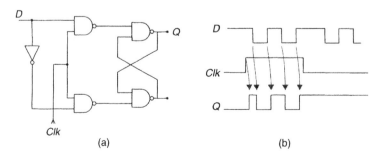

Figure 2.2. (a) Clocked D-latch; (b) timing diagram of clocked D-latch.

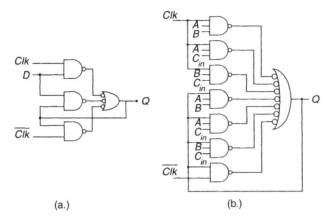

Figure 2.3. (a) Basic Earl's Latch; (b) implementing the Carry function.

In order to avoid the transparency feature introduced by the latch, an arrangement is made in which two latches are clocked back-to-back with two nonoverlapping phases of the clock. In such an arrangement the first latch serves as a Master by receiving the values from the data input, D, and passing them to the Slave latch, which simply follows the Master. This is known as a Master–Slave latch (MSL) (or L_1–L_2 latch, in IBM), as shown in Fig. 2.4. This is not a flip-flop, as we will explain later in this book. A very common VLSI implementation of MSL is the Transmission-Gate MSL, used in PowerPC (Gerosa et al. 1994), as shown in Fig. 2.4d.

In a M–S arrangement, the slave latch can have two or more masters acting as an internal multiplexer with storage capabilities. The first master is used for capturing data input, while the second master has other uses and can be clocked with a separate clock. One arrangement that utilizes two masters is the well-known IBM level-sensitive scan design (LSSD 1985) shown in Fig. 2.5.

In systems designed with LSSD compliance (Fig. 2.5), the system is clocked with clocks C and B during the normal operation and the storage elements act as standard MSLs. However, all storage elements in the system are interconnected by the alternate master in a long shift register. The input and the output of this shift register are routed to the external pins. In the test mode, the system is clocked with the A and B clocks, which act as a long shift register so that the state of the machine can be *scanned out* of the system and/or a new state *scanned in*. This greatly enhances the *controllability* and *observability* of the internal nodes of the system. LSSD is a mandated standard practice of all IBM designs, and it has become known in the industry as *boundary scan* (IEEE Standard 1149).

2.1.1. True-Single-Phase-Clock Latch

The true-single-phase-clock (TSPC) latch (Fig. 2.6), developed by Yuan and Svensson (1989), is a fast and simple structure that uses a single-phase clock. This latch was constructed by merging CMOS Domino and CMOS NORA

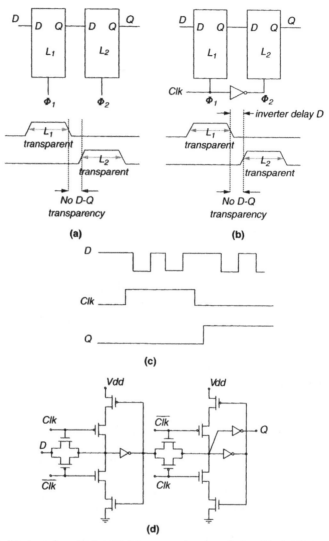

Figure 2.4. Master–slave latch with (a) nonoverlapping clocks; (b) single external clock; (c) timing diagram; (d) as used in PowerPC 603 (Gerosa, JSSC 12/94), Copyright © 1994 IEEE.

logic (Goncalves and De Man 1983). During the active clock ($Clk = 1$), CMOS Domino evaluates the input in a monotonic fashion (only a transition from logic 0 to 1 is possible) while NORA logic precharges. Alternatively, during the inactive clock ($Clk = 0$), Domino is being precharged (and so is nontransparent) while NORA is evaluating its input. The combination of NORA and Domino logic stages results in a nontransparent MSL that only requires a single clock. Hence the name given to it was true-single-phase-clock M–S latch. The clock system based on the TSPC M–S latch is described in Afghahi and Svensson (1990).

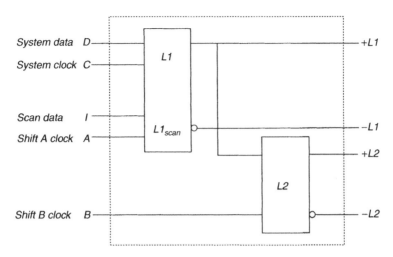

Figure 2.5. IBM LSSD compatible storage element.

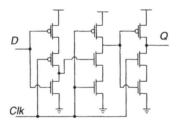

Figure 2.6. True-single-phase-clock (TSPC) M–S latch introduced by Yuan and Svensson (1989), Copyright © 1989 IEEE.

Operation of the TSPC M–S latch is illustrated in Fig. 2.7. When $Clk = 0$, the first inversion stage, L_1, is transparent and the second half, L_2, of the TSPC is precharged. Thus, at the end of the half-cycle, during which $Clk = 0$, the input D is present at the input of the Domino block as its complement, $\overline{D}$. When the clock switches to logic 1 ($Clk = 1$), Domino logic evaluates and the output, $\overline{Q}$, either stays at logic 0 or makes the transition from 0 to 1, depending on the sampled input value, $\overline{D}$. This transition cannot be reversed until the next clock cycle. In effect the first inverter connected to the input acts as a master latch, while the second (Domino) stage acts as a slave latch. The transfer from the master latch to the slave latch occurs as the clock changes its value from logic 0 to logic 1. Thus, the TSPC MSL behaves as a leading-edge triggered flip-flop. It is also frequently called a flip-flop, though by the nature of TSPC operation, this classification is incorrect.

Due to its simplicity and speed, the TSPC MSL was a very popular way of implementing a clocked storage element. However, the TSPC MSL was sensitive to glitches created by the clock edges. One of these glitches occurs on the output with a logic value of 1, while the input is receiving $D = 0$.

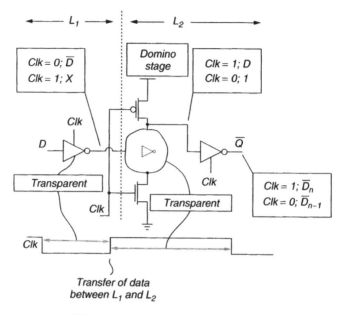

Figure 2.7. TSPC M–S latch operation.

2.1.2. Pulse Register Single Latch

Because of the high cost of the M–S latch design and the potential signal-race hazards introduced by the single-latch design, an idea for a single-latch design clocked by locally generated short pulses evolved. The idea is to make the clock pulse very short, and thus reduce the time window during which the latch is transparent. However, there is a possibility that a "short path" may be captured during the same clock. Given that the clock pulse is short, the chance of this hazard happening is reduced, and it is also possible to pad the logic (add inverters) in those paths so that they would not be a problem. Such a short clock pulse cannot be distributed globally because the clock distribution network would absorb it. There is an additional danger, because due to the process variations, the duration of that clock pulse will vary locally on the chip, as well as from chip to chip. In order to mitigate these problems, the pulse clock is generated locally, and it usually drives a register consisting of several such single latches that are physically located very close to each other. This method would lose its advantages of simplicity and low power if every single latch would require separate clock generator, as seen in Fig. 2.8a and 2.8b (Kozu et al. 1996).

The clock produced by the local clock generator must be wide enough to enable the latch to capture its data. At the same time, it must be sufficiently short to minimize the possibility of critical race. Those conflicting requirements make the use of this single-latch design hazardous by reducing the robustness and reliability of the design. Nevertheless, this design has been used because of the critical need to reduce the high costs imposed by the clocked storage elements.

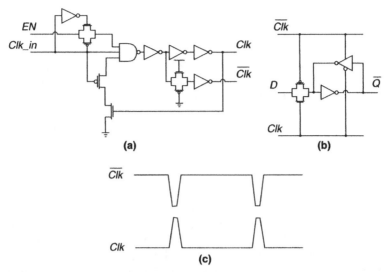

Figure 2.8. Pulse latch: (a) local clock generator; (b) single latch (Kozu et al. 1996); (c) clock signals, Copyright © 1996 IEEE.

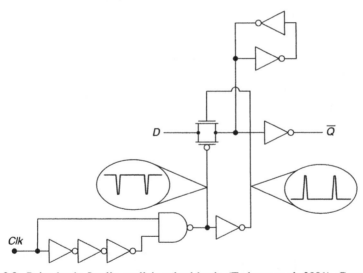

Figure 2.9. Pulse latch: Intel's explicit pulsed latch. (Tschanz et al. 2001), Copyright © 2001 IEEE.

Intel's version of the pulsed latch is shown in Fig. 2.9. One benefit of this design is low power consumption due to the common clock signal generator and a simple structure of the latch. In order to obtain the desired short clock pulse, the pulse generator used in Intel's pulsed latch uses the principle of reconvergent fan-out with nonequal parity of inversion.

2.2. FLIP-FLOP

The main feature of the flip-flop is that the process of capturing data is related to the transition of the clock (from 0 to 1 or from 1 to 0), thus the flip-flop is *not transparent*. Therefore flip-flop-based systems are easier to model, and the timing tools find flip-flop-based systems simpler and less problematic to analyze. The precise point in time when data are captured is determined by the clock event designated as either the leading or trailing edge of the clock. In other words, the transition of the clock from logic 0 to logic 1 causes data to be captured (it is the 1-to-0 transition in the trailing edge-triggered the flip-flop). In general, the flip-flop is not transparent, since it is assumed that the clock transition is almost instantaneous. As we will see later, even the flip-flop can have a very small period of transparency associated with the narrow time window during which the clock changes, as will be discussed later. In general, we treat the flip-flop as a nontransparent clocked storage element. Given that the triggering mechanism of a flip-flop is the transition of the clock signal, there are several ways of deriving the flip-flop structure. To better understand its functionality, it helps to look at an early version of a flip-flop, shown in Fig. 2.10, that was used in early computers and digital systems (see Siewiorek et al. 1982). The pulse, which causes the change, is derived from the *triggering* signal (also referred to as trigger) by using a simple differentiator consisting of a capacitor C and resistor R. One can also understand a glitch introduced by the flip-flop. If the triggering signal transition is slow, a pulse derived in this way may not be capable of triggering the flip-flop. On the other hand, even a small glitch on the trigger line can cause a false triggering.

To further our understanding of the flip-flop, it is helpful to start making the distinction between the flip-flop and the latch-based CSE.

The flip-flop and the latch operate on different principles. While the latch is "level-sensitive" meaning it is reacting on the level (logical value) of the clock signal, the flip-flop is "edge sensitive," meaning that the mechanism of capturing

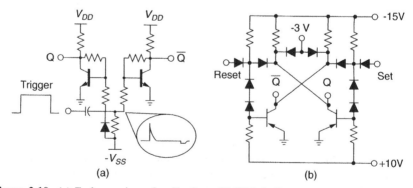

Figure 2.10. (a) Early version of a flip-flop; (b) PDP-8 direct set–reset sequential element.

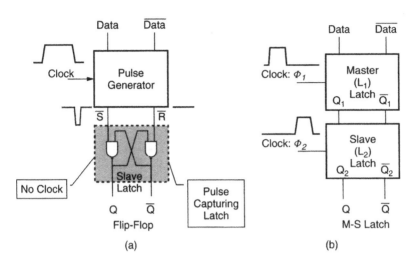

Figure 2.11. (a) General flip-flop structure; (b) general M–S latch structure.

the data value on its input is related to the changes in the clock signal. Level sensitivity implies that the latch captures the data during the entire period of time when the clock is active (logic 1), which means the latch is transparent. The two are designed from a different set of requirements, and so consist of inherently different circuit topologies.

The general structure of the flip-flop is shown in Fig. 2.11a. The difference between a flip-flop structure and the M–S latch, shown in Fig. 2.11b, is as follows:

A flip-flop consists of two stages: a pulse generator (PG) and a pulse capturing latch (CL). The PG generates a negative pulse on either the $\overline{S}$ or $\overline{R}$ lines, which are normally held at logic 1. The resulting pulse is a function of data and clock signals, and should be of sufficient duration to be captured in the pulse CL. The duration of the pulse produced by the PG stage can be as long as half the clock period, or it can be as short as one inverter delay.

On the other hand, the MSL consists of two identical clocked latches and its nontransparency is achieved by phasing clocks C_1 and C_2, which are clocking the master latch, L_1, and the slave latch, L_2.

In spite of the different topologies for the flip-flop and MSL, it may seem that because their outward appearance is the same, there is no difference between the two. In addition, the reader may believe that the distinction between the flip-flop and MSL is artificial and only of academic interest. Figure 2.12a shows the black-box view of the flip-flop and MSL. It appears that the MSL behaves identically to the trailing-edge-triggered flip-flop, so there is no apparent difference between the two. However, if the rise (or fall) time of the triggering edge of the clock increases, there will be a time at which the flip-flop will fail. This is illustrated in Fig. 2.12b, where the leading-edge-triggered flip-flop and MSL are compared. The MSL will continue to operate correctly, because the capturing mechanism

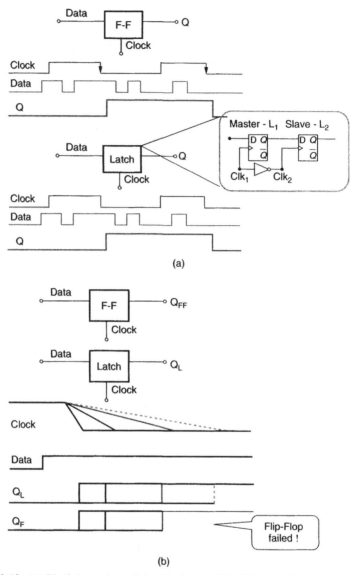

Figure 2.12. (a) Black-box view of the flip-flop and M–S latch; (b) experiment causing the flip-flop to fail while the M–S latch is still operational.

of both Master and Slave latches is related to the clock level, not to the rate of change. However, there are several reasons why the flip-flop may fail:

1. Degradation of the rate at which the clock signal changes (clock edge degradation) can diminish the level and duration of the internally produced pulse that sets the CL.

2. The difference in threshold levels of the gates used (due to the process variation) can cause the timing difference to behave differently than expected, resulting in no pulse being produced.

3. Any spurious glitch on the clock signal can cause false triggering of the flip-flop.

The experiment shown in Fig. 2.12 demonstrates the difference between the flip-flop and the MSL. This sensitivity of the flip-flop to the rate of the triggering edge makes the flip-flop potentially hazardous and a reliability problem in the systems where we cannot guarantee that the clock signal will suffer no degradations. This is particularly important where clock-edge degradation and noise on the clock signal lines are concerned.

Purely digital implementation of a flip-flop is far more intricate. For that analysis, the reader is referred to the commonly used SN7474 D-type flip-flop introduced by Texas Instruments and shown in Fig. 2.13 (Texas Instruments 1984). The analysis of the SN7474 flip-flop is particularly interesting because even a brief analysis reveals that the operation of this particular flip-flop is based on the races in time inside the first stage of this flip-flop.

The PG stage of the SN7474 is shown in Fig. 2.14, which may be helpful in the analysis of its operations and its failure modes. In order to behave as a flip-flop (to be sensitive to the change in the raising edge of the clock), an intricate race is introduced in the PG block that prevents any change on the $\overline{S}$ and $\overline{R}$ lines after the clock has moved from logic 0 to logic 1 (Oklobdzija 1999). Figure 2.14a is used to aid in the analysis of the PG block of the SN7474. Delay mismatch that can occur due to the process variations may result in this flip-flop malfunctioning, as shown in Fig. 2.14b. In the particular example shown, a race occurred between the $\overline{S}$ and $\overline{R}$ signals, which should be both stable at 1 after $\overline{S}$

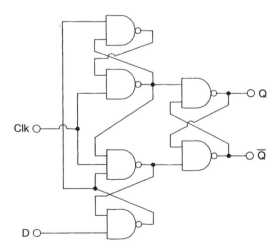

Figure 2.13. Texas Instruments SN7474 flip-flop.

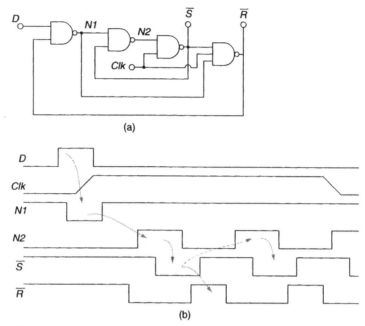

(a)

(b)

Figure 2.14. (a) Pulse generator block of SN7474; (b) malfunctioning due to a gate-delay mismatch.

has made a brief transition to 0 following the capture of $D = 1$ on the raising edge of the clock. Signal $\overline{R}$ should have stayed at 1 the whole time. In this particular case, the large difference in delay (due to the process variations) from one gate to another was the cause of this race.

The relationship of $\overline{S}$ and $\overline{R}$ signals with respect to the Data (D) and Clock (Clk) signals can be expressed as

$$S_n = Clk\,\overline{R}(D + S) \quad \text{and} \quad R_n = Clk\,\overline{S}(\overline{D} + R) \tag{2.1}$$

These expressions were derived strictly from the logic topology of the SN7474 flip-flop, shown in Fig. 2.13. The expressions for the next value of the set signal, S_n (as well as reset signal, R_n), provide a quick and simple insight into the functioning of the PG block of this flip-flop. Simply stated in words, the equation for S_n tells us: The next state of this flip-flop will be set to 1 only at the time the clock becomes 1 (rising edge of the clock), the data at the input are 1, and the flip-flop is in the steady state (both S and R are 0). The moment the flip-flop is set ($S = 1$, $R = 0$), no further change in data input can affect the flip-flop state, data input will be "locked" to $S_n = 1$ by $(D + S) = 1$, regardless of D, and reset R_n would be disabled (by $S = 1$). This assures the edge sensitivity, that is after the transition of the clock and setting the S_n or R_n signal to the desired state, the flip-flop is locked. No changes can occur until the clock transition to 0 (making both $S = R = 0$), thus enabling the flip-flop to receive new data.

Flip-Flop Derivation Given the set of specifications that describe the flip-flop property given earlier, we can undertake the process of deriving the logic equations for the flip-flop. We know that the flip-flop consists of: (1) a PG and (2) a pulse CL (Fig. 2.11a). The CL is a simple cross-coupled NAND (or NOR), set-reset (S–R) latch. We will see later how this CL can be designed in a very efficient way (Oklobdzija and Stojanovic 2001). The PG stage is specified by its expected behavior. The value of the PG outputs, S and R, after the clock makes its transition from 0 to 1 (triggering edge) is a function of the Clock, Data and the previous values of S and R. A description of S_n is given in the previous section. For clarity, we will repeat it specifically for the required next value of the S_n signal: The next state of the flip-flop should be set to 1 only at the time the clock becomes 1 (triggering edge of the clock), the data at the input are 1, and the flip-flop is in the steady state (both S and R are 0). The moment the flip-flop is set ($S = 1$, $R = 0$) no further change in data input can affect the flip-flop state. Therefore, S_n should become 1 when the clock becomes 1 and data is 1. When this event occurs, S_n stays at 1 and it cannot revert back to 0, even if the data signal changes back to 0.

It is quite simple to show these functional specifications on a Karnaugh map, as shown in Fig. 2.15. Now we can derive logic equations from the functional specifications given in the Karnaugh map; these equations are equivalent to the ones shown in Eq. (2.1).

If we use the equations obtained this way to construct a PG of the flip-flop, it will result in the circuit topology shown in Fig. 2.16c. Combining the PG stage obtained with the improved second-stage CL invented by Stojanovic (Oklobdzija and Stojanovic 2001) results in a superior flip-flop that was later implemented and further enhanced by Nikolic et al. (1999). This flip-flop is in the leading group of high-performance flip-flops in terms of speed and energy delay product.

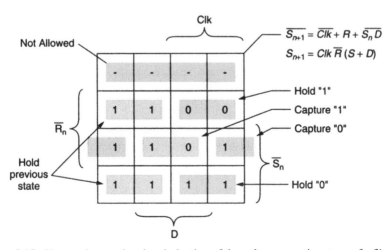

Figure 2.15. Karnaugh map showing derivation of the pulse-generating stage of a flip-flop (only the S_n signal is shown).

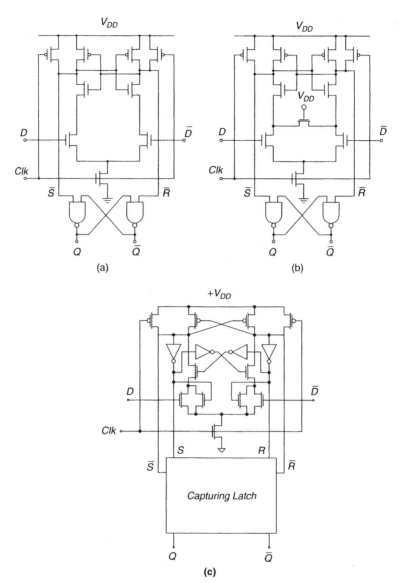

Figure 2.16. (a) Pulse generator stage of the sense-amplifier flip-flop. (Madden and Bowhill, (1990); (b) Improvement for floating nodes. (Dobberpuhl 1997; Montanaro et al. 1997) (c) Pulse generator stage improvement by proper design. (Nikolic and Oklobdzija 1999). Copyright © 1990, 1997 IEEE.

It is interesting to note that engineers had to make several attempts before they found the right circuit topology for this flip-flop. The flip-flop used in the third generation of Digital Equipment Corporations 600-MHz Alpha (Gronowski et al. 1998) processor is a version of the flip-flop introduced by Madden and Bowhill, which was based on the static memory cell design (Madden and Bowhill 1990). This particular flip-flop is known as a sense-amplifier flip-flop (SAFF) (Matsui et al. 1994). The development of the PG block of this flip-flop is illustrated in Fig. 2.16a–2.16c.

The behavior of the SN7474 flip-flop is identical to that of Alpha's SAFF. When setting the flip-flop, both of them hold the $\overline{S}$ (or $\overline{R}$) line at logic 0 for the duration of the clock active (logic 1) and reset them to logic 1 once the clock returns to 0 (inactive state).

One of the objectives of this book is to clarify the confusion that exists in understanding and properly classifying various types of clocked storage elements. In the next section we will show another way (used in practice) to create a flip-flop. In the SN7474, disabling the D input is done after the short delay necessary to set S (or R) to the next value, thus achieving the *edge property*. This short delay is essential and cannot be avoided. It is reflected in the parameters of the setup and hold times of the flip-flop, which will be discussed later in the book.

2.2.1. Time Window-Based Flip-Flops

Derivation   Digital circuits are based on discrete events. Not only are the logic signals a set of discrete voltage levels, but the time is also based on either the clock (leading or trailing edge) or some other finite delay based on the signal propagation through one or more of the logic elements. Determining when to shut the flip-flop off is also based on a discrete time event with reference to the clock, or one or more inverter or gate delay units following the transition of the clock. This method is illustrated in Fig. 2.17, where one buffer delay serves as a time reference for shutting the flip-flop off. Thus, the clock edge is created to last during a time interval (window) from Clk to Clk_1, during which the flip-flop may be transparent. When $D = 1$ and $Clk = 1$, S_{n+1} changes to 0 and immediately back to 1 as soon as $Clk_1 = 1$. At this point any change in D has no effect on S_{n+1}, because any further input transition is blocked. This describes the following flip-flop property: $\overline{S_{n+1}} = \overline{Clk} + \overline{Clk_1}D + Clk_1\overline{S}$, which means that $S_{n+1} = 0$ only for the short period of time until $Clk_1 = 1$; afterwards the state is maintained by the term $Clk_1\overline{S}$, while data can have no effect because $\overline{Clk_1}D = 0$. Thus, nontransparency is assured after the clock edge.

The usual technique for generating the time reference is to create a short pulse using the property of *reconvergent fan-outs with nonequal parities of inversion*. This arrangement, which uses the clock signal and three inverters with both paths reconverging as inputs of a NAND gate, is shown in Fig. 2.9. The trailing edge of this pulse is used as a time reference for shutting off the flip-flop. Depending on the particular implementation, a short transparency window can be introduced. This transparency window has been a source of confusion in classifying these flip-flops. One example is a flip-flop introduced under the name "Hybrid-Latch Flip-Flop" (HLFF). The existence of a short transparency window caused its inventor to treat it as a latch, but since its behavior was not that of a latch, it was given its dual name (Partovi et al. 1996). The HLFF is shown in Fig. 2.18.

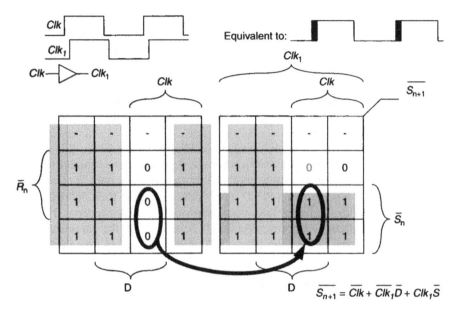

Figure 2.17. Method of creating the time reference points for opening and shutting the flip-flop.

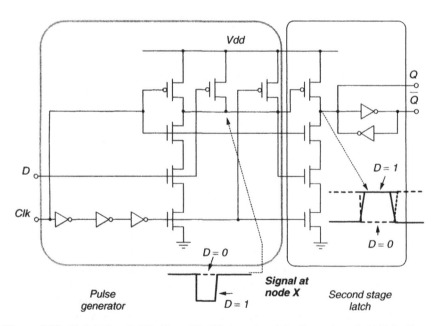

Figure 2.18. Hybrid-Latch Flip-Flop (HLFF) introduced by Partovi et al. (1996). Copyright © 1996 IEEE.

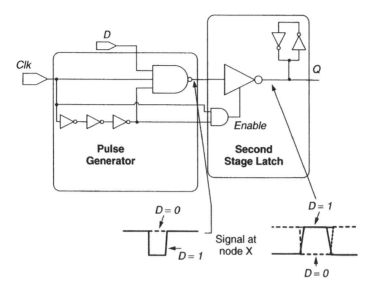

Figure 2.19. Logic representation of Partovi's flip-flop (HLFF).

Detailed analysis shows that the number of transistors has been reduced from the original specifications, which resulted in imperfections in the behavior of this flip-flop. A logic representation of this flip-flop shows two NAND gates connected in series (Fig. 2.19). The first NAND gate creates the pulse if $D = 1$. Here, the data signal serves as a *pulse enabler* or *pulse inhibitor*, depending on the value of D.

The problem with this structure is that its second stage is incomplete, which serves as a clockless CL. In order to avoid an excessive number of p-MOS transistors and obtain latch functionality, the second NAND gate is not fully implemented and its output node floats when the output node X (from the first NAND) is at logic 1 after the pulse has ended. In essence, this node (X) represents the $\overline{S}$ signal from the pulse generator. The absence of the $\overline{R}$ signal, due to the single-ended implementation of this flip-flop, hinders the ability to completely realize the flip-flop function. This is not a case of complete SAFF implementation (Nikolic and Oklobdzija 1999). The floating output node of the HLFF is susceptible to glitches and even the slightest mismatch of clock signals between the first and second stages. When data input $D = 1$, the leading edge of the clock sets $X = 0$ (precharged node), but only after some propagation delay caused by the time it takes to make $X = 0$ (set operation). All three inputs of the second NAND gate will be at 1 for a short time after the leading edge of the clock. This will cause a glitch in the output, a problem that is inherent in the HLFF structure.

Figure 2.20 shows a systematic approach to deriving a single-ended flip-flop. The flip-flop shown in Fig. 2.20 has three time reference points: (1) leading edge of the *Clk* signal; (2) trailing edge of the *Clk* signal after passing through three

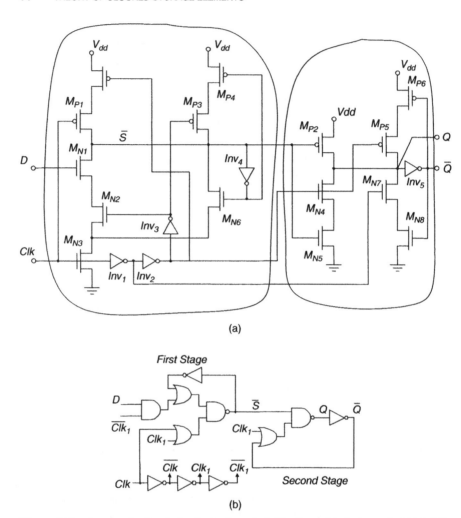

Figure 2.20. Systematically derived single-ended flip-flop: (Nedovic and Oklobdzija 2000a) (a) circuit diagram; (b) logic representation. Copyright © 2000 IEEE.

inverters, Inv_{1-3}; (3) leading edge of the Clk signal after passing through two inverters, Inv_{1-2}. The clock signal Clk is designated Clk_1 after two inversions, and Clk_1 after three inversions. The logic representation of this flip-flop is shown in Fig. 2.20b.

The model for this leading edge triggered flip-flop uses three time reference points. Equations (2.2)–(2.4) describe the behavior of this flip-flop.

Pull-down path is the implementation of:

$$\overline{S} = \overline{Clk(D + \overline{Clk_1} + S)} \qquad (2.2)$$

The pull-up path is implemented using:

$$\overline{S} = \overline{(Clk + Clk_1)(\overline{Clk_1} + S)} \tag{2.3}$$

This enhances performance a little (by reducing the capacitance on node $\overline{S}$) without significant degradation in reliability. The second stage (capturing latch) is implemented as:

$$Q = \overline{\overline{S}(Clk_1 + \overline{Q})} \tag{2.4}$$

Clock signal Clk_1 delays capture of the value on $\overline{S}$ until node $\overline{S}$ stabilizes. This eliminates the hazard encountered in the HLFF (Partovi et al. 1996) and SDFF flip-flops (Klass et al. 1998). In addition, a systematically derived flip-flop (Nedovic and Oklobdzija 2000a) exhibits better speed when compared to the HLFF and SDFF.

CHAPTER 3

TIMING AND ENERGY PARAMETERS

This chapter deals with the timing and energy parameters of CSEs. We discuss the various definitions of timing parameters and provide insight into energy consumption in clocked storage elements.

3.1. TIMING PARAMETERS

Latches and flip-flops have different timing characteristics in general. However, it is possible to establish some common parameters for both. These parameters are based on timing relations between data and clock inputs that ensure correct circuit operation. We define basic timing parameters using a flip-flop and extend the analysis to a latch.

3.1.1. Clock-to-Output Delay, t_{CQ}

The clock-to-output delay is the delay measured from the clock triggering edge to the output. It is a function of the arrival of data and clock signals, the slope of these signals, the supply voltage, temperature, process parameters, and the output load.

Basic timing diagrams of flip-flops are illustrated in Fig. 3.1. The flip-flop samples data, D, at the clock triggering edge (leading edge in this example) and generates the appropriate output after the propagation delay, $t_{CQ,LH}$ if output undergoes a 0–1 transition or $t_{CQ,HL}$ if output undergoes a 1–0 transition. The transitions occur between two consecutive clock edges, provided there is no violation of timing constraints between the data and clock inputs. Fundamental timing constraints between data and clock inputs are quantified with setup time,

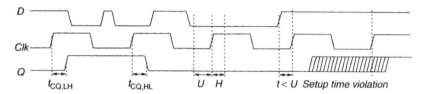

Figure 3.1. Basic timing diagrams in flip-flops.

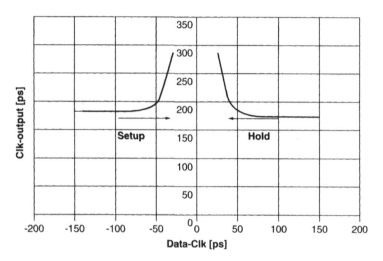

Figure 3.2. Setup and hold time behavior as a function of clock-to-output delay.

U, and hold time, H. Data have to be stable at least setup time before and hold time after the active clock edge, as illustrated in Fig. 3.1.

Having defined basic timing relationships related to the setup and hold times, question about the failure mechanism of the clocked storage element remains. If we establish an experiment in which we set the data arrival closer to the clock, we see that at first the Clk-Q delay of the storage element will start to increase before the capturing mechanism fails. Something similar happens at the other end when the next data arrival gradually approaches the current clock edge. This is not an abrupt process, as the definition of the setup and hold times implies. This behavior is shown in Fig. 3.2. Obviously we do not want to allow the data to come too close to the failing region for fear that we may have an unreliable design. However, keeping the data too far from the failing region takes away precious cycle time, which impacts the performance negatively. This creates a need for more precise definitions of the setup and hold times.

3.1.2. Setup Time, *U*

Although there are various definitions of the setup time, they all relate to the same fundamental mechanism–degradation of Clk-Q delay due to a change in the relative arrival time of data and clock signals, described in the previous section. The value of the setup time depends on the distribution of the internal clock signal inside a CSE, and can be both positive and negative. MSL typically have a positive setup time, while pulsed latches and flip-flops usually have a negative setup time, often accompanied by the soft-edge property (Partovi et al. 1996).

To fully understand the impact of the setup time on the overall system cycle time, one needs to consider the setup time in conjunction with the Clk-Q delay. The sum of the two is the only true measure of the CSE delay relative to the overall cycle time. Therefore, the setup time that is based on minimum an achievable data-to-output delay is the *optimum* setup time from the perspective of the impact of the CSE delay on cycle time (Stojanovic and Oklobdzija 1999). This is illustrated in Fig. 3.3, which shows that when D arrives later relative to Clk, D-Q delay initially decreases because D still arrived more than one setup time before Clk, which is still early enough so that no significant increase in Clk-Q delay can be observed. When D is further delayed, then at one point the increase in Clk-Q grows larger than the absolute decrease in the D-Clk delay, so the overall D-Q delay starts increasing, until the capturing mechanism fails. Therefore, minimum D-Q delay exists. Data arrival resulting in the minimum D-Q delay would therefore be the optimal setup time, corresponding to a $45°$ slope on the Clk-Q characteristics (Fig. 3.3). Optimally tuning the data arrival close to this point is a hard task in general, because the arrival times of D and Clk are not easy to control due to variations in logic delay and clock skew. However, if the CSEs are spatially close, which is the case in critical paths, these variations would be reduced, and the designer would be able to fine-tune the logic and CSE delays to achieve almost optimal clock frequency. This approach is applicable to custom, high-performance designs, where achieving peak performance is the ultimate goal.

Another method for quantifying the setup time is based solely on the Clk-Q delay. It is defined as the D-Clk delay that corresponds to some specified increase in the Clk-Q delay, relative to the nominal Clk-Q delay. The nominal Clk-Q delay is defined as the Clk-Q delay when D arrives early before the Clk

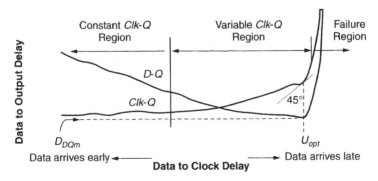

Figure 3.3. Setup time behavior as a function of data-to-output delay.

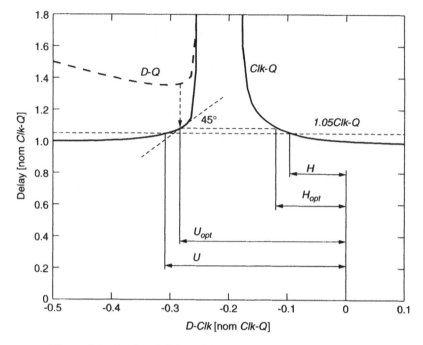

Figure 3.4. $D-Q$ and Clk-Q delay as a function of D-Clk offset.

so that there is no degradation in the Clk-Q delay. This method for obtaining U is illustrated in Fig. 3.4, where a 5% increase in the Clk-Q delay is used (Markovic et al. 2001). The setup time obtained using this approach does not necessarily coincide with the min $D-Q$ setup time, which makes it suitable for designs where performance is not the primary concern. This is typically the case in standard cell-based synthesized systems.

Neither of these definitions is preferred over the other. Our goal was to present both approaches and hint at their applicability. Either of the two definitions presented here can be used, depending on the application, available design tools, and preference of the designer. Additionally, the designer should always keep the worst-case reliability conditions in mind and back off in time in order to tolerate process–voltage–temperature variations, and be sure that the CSE will operate correctly, because the failure region does not occur until after the setup time point.

3.1.3. Hold Time, *H*

While the setup time can be obtained either by using $D-Q$ or the Clk-Q delay, there is no such ambiguity in the definition of the hold time. This is simply because the $D-Q$ delay does not capture the region of the hold time violation, as shown in Fig. 3.4. Instead, the hold time is obtained from the Clk-Q

versus *D-Clk* characteristics. It is typically determined to be the *D-Clk* offset corresponding to some specified increment in *Clk-Q* delay from its nominal value. As an illustration, Fig. 3.4 defines hold time as a 5% increase in the *Clk-Q* delay. In addition, Fig. 3.4 also defines optimal hold time, H_{opt}, as the *Clk-Q* delay increment corresponding to the optimal setup time, U_{opt}.

The hold time is equally important in both high-performance and low-energy designs. It relates to early data arrival, where timing violations due to critical races can occur. This directly translates into the clock-skew budget.

The sum of the setup time and hold time defines a minimal data width, the time during which data must remain stable. Setup and hold times are different in flip-flops and latches.

Setup and Hold Times in Flip-Flops For purposes, of illustration, setup time, U, hold time, H, sampling window, and clock width, w, for a flip-flop are shown in Fig. 3.5. Setup and hold times are therefore related to the triggering edge of the clock, in this case, the leading edge.

Setup and Hold Times in Latches The situation with the latch is different, as illustrated in Fig. 3.6. The setup time for the latch starts from the trailing edge of the clock signal, because closing the latch is an action that would capture the last data present in the latch. In addition, there are two delay times defined t_{CQ} (as in the flip-flop) and t_{DQ} because of the two possible scenarios: (1) data being present and waiting for the clock to open the latch, and (2) data arriving while the latch is open.

It would be appropriate to observe that the failure mode of the flip-flop does not necessarily follow the failure mode of the latch as a result of the violations of the setup or hold times. Depending on the flip-flop implementation, violation of the setup or hold times can lead to oscillations in the pulse generator stage of the flip-flop, as discussed in Chapter 2. As a result, once the oscillation occurs the

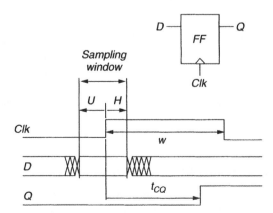

Figure 3.5. Setup time, hold time, sampling window, and clock width in a flip-flop.

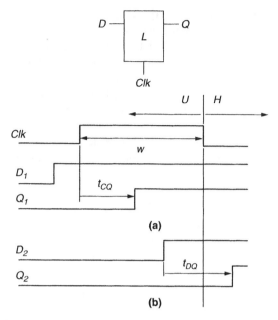

Figure 3.6. Latch: setup and hold time: (a) early data D_1 arrival; (b) late data D_2 arrival.

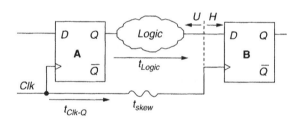

Figure 3.7. Illustration of a data path. (Markovic et al. 2001), Copyright © 2001 IEEE.

output value, Q, cannot be predicted. These oscillations in the flip-flop usually occur abruptly, as opposed to the more gradual delay increase encountered with the latch. Therefore one needs to be more careful with the flip-flop than with the latch-based design.

Having defined setup and hold times, the next step is to illustrate their significance in a true data-path design, as shown in Fig. 3.7. In order to accomplish this, we introduce the concept of early and late data arrival.

3.1.4. Late Data Arrival and Time Borrowing

From the graph in Fig. 3.3 we see that in spite of Clk-Q delay increasing, we are still gaining in terms of $D-Q$ delay, because the time taken from the cycle is reduced. In other words, the increase in the storage element delay is still smaller

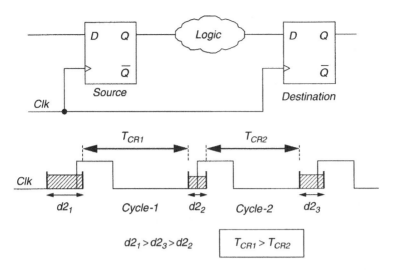

Figure 3.8. Time Borrowing in a pipelined design.

than the amount of time for which data is delayed, thus allowing more time in the cycle for the useful logic operation. Thus, we are encountering some new phenomena: *time borrowing*, *cycle stealing*, and *slack passing*. We will use the term time borrowing later in the text. In order to understand the full effects of delayed data arrival, we have to consider a pipelined design where the data captured in the first clock cycle is used as input data in the next clock cycle, as shown in Fig. 3.8.

As can be seen in Fig. 3.8, the data-to-output time window, $d2$, moves around the time axes. The parameter $d2$ is defined by the latest data arrival and by valid CSE output. As the data arrive closer to the clock, the size of the $d2$ shrinks (up to the optimal point). The data in the next cycle will then arrive later compared to the case where the data in the previous cycle were ready well ahead of the setup time. The amount of time for which the T_{CR1} was augmented did not come for free. It was simply taken away (stolen or borrowed) from the next cycle T_{CR2}. As a result of late data arrival in Cycle-1, there is less time available in Cycle-2. Thus the boundary between the pipeline stages is somewhat flexible. This feature not only helps accommodate a certain amount of imbalance between the critical paths in the various pipeline stages, but it helps in absorbing the clock skew and jitter. Thus, time borrowing is one of the most important characteristics of high-speed digital systems.

3.1.5. Early Data Arrival and Internal Race Immunity

The maximum clock skew that a system can tolerate is determined by the clock storage elements. To quantify this timing measurement, internal race immunity R is introduced. If the Clk-Q delay of storage element A is shorter than the hold time of storage element B in Fig. 3.7, and there is no logic in-between, a race condition can occur. In other words, there is a *minimum delay restriction* on the *Clk-Q* delay given by Eq. (3.1). Internal race immunity, R, of a clocked storage element is given by Eq. (3.2):

$$t_{Clk-Q} \geqslant H + t_{skew} \tag{3.1}$$

$$R = t_{Clk-Q} - H \tag{3.2}$$

Internal race immunity, R, of a clocked storage element is the difference between its *Clk-Q* delay and hold time, H. If it is greater than the maximal clock skew, we do not have to worry about minimal delay restrictions. The internal race immunity is a helpful measurement that aids in the analysis of timing failures due to short paths (races). In addition, it relates to the maximum clock skew a CSE can tolerate.

3.1.6. Minimum Data Pulse Width

The minimum width of the data pulse is the minimum time during which data are required to be stable to ensure correct operation of a clocked storage element. It defines the sampling window, and it is approximately equal to the sum of the setup and hold times. The minimum data pulse width over a range of supply voltages is illustrated in Fig. 3.9. The minimum pulse width widens with scaling of the supply voltage, meaning that an extra margin has to be included to achieve

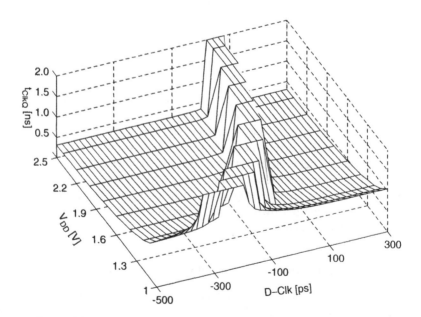

Figure 3.9. Impact of supply voltage on the minimum data pulse width.

sufficient robustness in the design. In addition, the relative arrival of D and Clk shifts with scaling the supply voltage. This is a direct consequence of the delay scaling.

3.2. ENERGY PARAMETERS

The battery life in portable devices is proportional to their energy consumption. In high-performance designs, energy consumption has a large impact on the design and may limit performance. It is therefore imperative to design the digital circuits, used in consumer products, that consume the minimum amount of energy for a given task. In order to accomplish this, designers need to understand energy consumption in the various circuits used in the implementation. In analysis of energy consumption in the clock subsystem, the designs should look specifically at the energy consumed in the clocked storage elements.

3.2.1. Components of Energy Consumption

Total energy consumption in a CSE during one clock period is obtained using Eq. (3.3):

$$E = \int_{t}^{t+T} V_{DD} \cdot i_{V_{DD}}(\tau) \cdot d\tau \qquad (3.3)$$

where t is the time point chosen in a way that includes all relevant transitions: arrival of new data, clock pulse, and output transition. This energy has four components: switching, short circuit, leakage and static energy, which are briefly explained in this section:

$$E = E_{\text{switching}} + E_{\text{shortcircuit}} + E_{\text{leakage}} + E_{\text{static}} \qquad (3.4)$$

Switching Energy   The switching component of energy is defined as:

$$E_{\text{switching}} = \sum_{i=1}^{N} \alpha_{0-1}(i) \cdot C_i \cdot V_{\text{swing}}(i) \cdot V_{DD} \qquad (3.5)$$

where N is the number of nodes in a circuit; C_i is the capacitance of the node i; $\alpha_{0-1}(i)$ is the probability that energy-consuming transition occurs at the node i; $V_{\text{swing}}(i)$ is voltage swing of the node i; and V_{DD} is the supply voltage. The switching component of energy is the main contributor to the overall energy consumption when the switching activity is high. Since the switching component contributes to the total energy the most, energy can be best reduced if each of the terms in the product expression is minimized. This becomes a simple design guideline for energy reduction in digital circuits where the switching component is the dominant component.

Short-Circuit Energy   The short-circuit component of energy arises from the short circuit (crowbar) current. The short-circuit current occurs when both pull-up

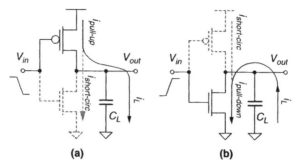

Figure 3.10. Short-circuit current in an inverter during: (a) pull-up; (b) pull-down operation.

and pull-down paths conduct current at the same time. To illustrate this, let's examine the simple case of a CMOS inverter, where due to the finite rise and fall times of the input waveform when $V_{Tn} < V_{in} < V_{DD} - |V_{Tp}|$, both the n-MOS and p-MOS transistors are on, which causes short-circuit energy consumption. During the pull-up operation, as shown in Fig. 3.10a, it is desirable that all pull-up current of the p-MOS transistor be delivered to C_L, in which case the current of the n-MOS transistor is short-circuit current. Similarly, for the pull-down operation, current of the p-MOS transistor represents the short-circuit current, as shown in Fig. 3.10b.

The short-circuit energy component is typically less than 10% of the total energy. However, it becomes much greater when the slope of the input signal is large in comparison with the slope of the output signal. The input signal slope defines the time interval during which both pull-up and pull-down devices are simultaneously on. In a well-designed system, the input and output slopes are balanced, with the output slope always being nearly as fast as the input slope, thus minimizing short-circuit current.

Leakage Energy   The leakage component of energy comes from two types of leakage currents: (1) reverse-bias diode leakage at the transistor drains, and (2) subthreshold leakage through the channel of a device that is off.

The diode leakage occurs when a transistor is off, and the drain-bulk or source-bulk diode is reverse-biased so that it conducts current. The leakage current of the reverse-biased diode is given by

$$I_{\text{leakage}} = I_{\text{sat}} \cdot \left(e^{V/V_t} - 1 \right) \tag{3.6}$$

where V is the diode voltage. When the diode is reverse-biased, its current is approximately equal to the reverse saturation current. This component is typically negligible compared to the subthreshold leakage component.

The subthreshold leakage component is due to carrier diffusion between the source and drain when the channel-to-substrate surface potential ϕ_S is given by $\phi_B < \phi_S < 2\phi_B$, which corresponds to the moderate inversion region, (Wolf

1995). The term ϕ_B represents the Fermi potential. The drain-to-source current in the subthreshold region is exponentially proportional to the gate-to-source overdrive, $V_{GS} - V_{TH}$, as given by Eq. (3.7):

$$I_{DS,\text{subthreshold}} \propto e^{(V_{GS} - V_{TH})/(n \cdot V_t)} \cdot \left(1 - e^{(-V_{DS}/V_t)}\right) \tag{3.7}$$

For $V_{DS} \gg V_t = kT/q$, the last term is approximately equal to 1, and I_{DS} is independent of V_{DS}, which typically happens for V_{DS} larger than 0.1 V (Chandrakasan 1994). This current is becoming increasingly important with the scaling of CMOS technology, because the subthreshold slope increases due to the increase in gate-to-drain overlap capacitance (Wolf 1995).

Energy consumption due to leakage currents is increasing in importance with the technology scaling. As an illustration, Fig. 3.11 shows projected off currents in four consecutive deep-submicron technology generations (Chandrakasan et al. 2001).

Assuming a 50% increase in the total transistor width per technology generation, the total leakage current would increase by about 7.5 times, corresponding to a 5 × increase in the total leakage power. Furthermore, for constant die size, the active power remains constant, indicating that the leakage power will soon become a significant portion of the total power consumption in modern microprocessors.

Static Energy   The component due to static currents appears in two cases: (1) reduced voltage levels driving CMOS circuits, and (2) circuits with DC current (e.g., pseudo-n-MOS circuits). Both of these cases rarely occur in CSE circuits.

3.2.2. Energy Breakdown

Understanding the energy breakdown inside clocked storage elements is the key to energy-efficient design. A system-level designer, for instance, may be particularly interested to know how much energy is consumed in the clocking of

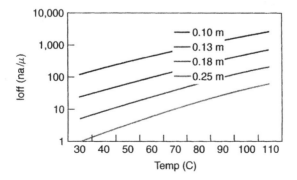

Figure 3.11. Projected off currents. (Chandrakasan et al. 2001), Copyright © 2001 IEEE.

a storage element with the objective of minimizing the energy of the clocking subsystem. Additionally, it is important to understand where the energy goes inside the storage element to be able to minimize other components of its energy consumption. This motivates the idea of finding the energy breakdown between (1) internal clocked nodes in storage elements, (2) internal nonclocked nodes in storage elements, (3) data and clock input load, and (4) output load.

Energy per transition measurement, which is introduced later in this section, aids in the calculation of energy breakdown in CSEs. Briefly, energy breakdown is the total energy consumed by a CSE during one of the four possible input data transitions. Table 3.1 summarizes energy breakdown in CSEs based on the energy-per-transition. The fields marked with both letters Y and N denote that the specific component is optional depending on the circuit structure. For example, energy E_{ext}, which is dissipated in charging the external load, is contained in E_{0-1} in noninverting CSEs, and in E_{1-0} in inverting CSEs.

Internal Clocking Energy   Clocking energy of a storage element is the energy consumed in its internal, clocked nodes. The total energy consumed by a CSE is consumed only in internal clock nodes when input data do not change. The clocking energy is therefore simply evaluated as E_{0-0} or E_{1-1}. In general, E_{0-0} and E_{1-1} are not equal, depending on the circuit structure. There are two cases to consider: (1) storage elements without precharge nodes, and (2) storage elements with precharge nodes.

Storage Elements without PreCharge Nodes   Clocking energy in storage elements that do not have precharge nodes is equal to the total energy consumed by the CSE. This is valid only when input data undergo 0–0 or 1–1 transition. Examples of this class of CSEs are the conventional MSL circuits (Suzuki et al. 1973; Gerosa et al. 1994). Provided that there are no precharge nodes, all energy is consumed only in the internal clocked nodes. For this reason, the energy consumed during 0–0 and 1–1 input transitions is the same:

$$E_{Clk} = E_{0-0} = E_{1-1} \qquad (3.8)$$

Storage Elements with Precharge Nodes   Storage elements with precharge nodes spend extra energy for precharging these nodes. Energy consumed in the internal clocked nodes is therefore not necessarily equal to the total CSE energy.

Table 3.1   Energy Breakdown in Clocked Storage Elements

	E_{0-0}	E_{0-1}	E_{1-0}	E_{1-1}
E_{Clk}	Y/N	Y	Y	Y/N
E_{int}	Y/N	Y	Y	Y/N
E_{ext}	N	Y/N	Y/N	N

The HLFF and SDFF are good examples of this class of circuits (Partovi et al. 1996; Klass 1998). In both of these designs, E_{0-0} and E_{1-1} are different because of the precharge energy required during $1-1$ input transition. The precharge node, however, remains precharged during the $0-0$ input transition, in which case all energy is consumed in the internal clocked nodes, Eq. (3.9). The difference between E_{1-1} and E_{0-0} is therefore equal to the energy consumed in the precharge node, Eq. (3.10):

$$E_{Clk} = E_{0-0} \tag{3.9}$$

$$E_{\text{precharge}} = E_{1-1} - E_{0-0} \tag{3.10}$$

In some CSEs with precharge nodes, however, it is not possible to separate the clocking energy from the precharge energy. This is the case, for example, in SAFFs (Matsui et al. 1994). Because of its differential nature, both $0-0$ and $1-1$ input transitions require that the internal nodes be precharged, in which case E_{0-0} and E_{1-1} represent the total energy consumed in both the precharge operation and in charging the internal clocked nodes.

Data and Clock Input Energy   Data input energy is simply the energy required to charge the capacitance of the data input. As shown in Fig. 3.12, this is the energy taken out of supply, $V_{DD\text{-}D}$, excluding the energy dissipated in driving the parasitic capacitance of the shaded inverter. Similarly, the external clock energy dissipated in driving the clock input of the CSE is the energy taken out of supply $V_{DD\text{-}Clk}$.

Energy in Internal Nonclocked Nodes   To calculate the energy consumed in the CSE's internal, nonclocked nodes, we need to identify the input data transition that results in energy consumption in these nodes. In the case of a noninverting CSE, this occurs during input transition $1-0$. The total energy, E_{1-0}, is then dissipated in charging the internal clocked and nonclocked nodes. To capture the average energy dissipated in the internal nodes, we need to consider the other transition as well, and subtract the clocking energy and energy dissipated in driving the load, C_L, Eq. (3.11). The energy consumed in charging the internal

Figure 3.12. CSE test setup.

precharged nodes can be either lumped into E_{int} or separately calculated as in Eq. (3.10).

$$E_{int} = \frac{E_{0-1} + E_{1-0}}{2} - E_{Clk} - E_{Load} \qquad (3.11)$$

Energy in Output Load   Energy is delivered to the external output load when output undergoes the 0–1 transition. This energy can be obtained as the energy drawn from the separate supply voltage (not shown in Fig. 3.12) that powers the output stage of the CSE.

3.2.3. Energy per Transition

A useful energy measurement for system designs is the *energy per transition.* The energy per transition is the total energy consumed in a CSE during one clock cycle for a specified input data transition (0–0, 0–1, 1–0, or 1–1) from the energy transition diagram; Fig. 3.13. This measurement is extremely crucial because it yields significant insight about circuit energy without the need for complex and intricate mathematical formulas. We can obtain this information empirically by running only one simulation to compute E_{0-0}, E_{0-1}, E_{1-0}, and E_{1-1}. Subsequently, these four values can be used to calculate the CSE energy consumption for any given input data pattern, as in Eq. (3.12), where p_{i-j} represents the probability of the $i-j$ input transition:

$$E_{average} = p_{0-0} \cdot E_{0-0} + p_{0-1} \cdot E_{0-1} + p_{1-0} \cdot E_{1-0} + p_{1-1} \cdot E_{1-1} \qquad (3.12)$$

By inspecting the node activity in a CSE for different input data transitions, the energy per transition can be utilized to obtain the energy breakdown between clocked nodes, internal nodes, and the external output load. This is a good basis for studying alternative circuit techniques that deal with internal clock gating. The energy breakdown information also offers valuable information about the tradeoffs associated with reduced clocking energy and the energy penalty incurred by the clock-gating logic, thus providing a better understanding of the optimization goals for the overall design.

3.2.4. Glitching Energy

In this section we analyze the energy consumed by dynamic hazards that are generated by the unintended transitions propagating from the fan-in gates,

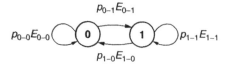

Figure 3.13. Energy transition diagram.

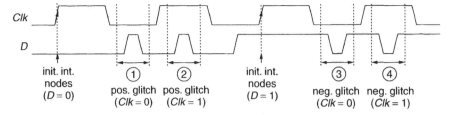

Figure 3.14. Types of glitches in CSEs.

often called *propagating glitches* (Hashimoto et al. 1998). Glitches produced by nonglitch transitions at the inputs, called *generated glitches*, are not covered in our discussion.

There are four types of glitches in CSEs, all of which can be represented as shown in Fig. 3.14. Average CSE glitching energy is determined by the glitching probability and the energy that the CSE consumes during glitching as in Eq. (3.13).

$$E_{\text{avg-glitch}} = \sum_{i=1}^{4} \beta_i \cdot E_{g_i} \tag{3.13}$$

In our analysis, we used simplified transition diagrams for regular and glitching transitions. More formal methods for calculating energy consumption due to regular transitions and glitches using state-transition diagrams can be found in (Zyuban and Kogge 1999).

3.3. INTERFACE WITH CLOCK NETWORK AND COMBINATIONAL LOGIC

The clocked storage element measurements described thus far considered the entire CSE, implicitly assuming that the data and clock inputs were supplied by drivers with sufficient drive strength. The input clock and data capacitances are important interface parameters for the clock network and logic design. The clock network designer and logic designer need to be aware of these capacitances in order to design circuits that drive storage elements.

3.3.1. Interface with Clock Network

The timing specifications of the clock distribution network that affect the clocked storage element parameters are *clock skew* and *clock slope*. The important energy parameter is the total load of the clock distribution network, which is defined by the input capacitance of the clock node and the number of storage elements on a chip.

An increase in clock slope results in degradation of the storage element performance, so the clock network designer has to know what slopes the storage elements can tolerate. This is especially important if flip-flops are used. The clock slope also affects the clock distribution network's energy consumption. If larger clock drivers with smaller fan-out are used, the clock edges are sharper and the storage element performance better, at the expense of an increase in energy consumption by the clock network. Optimal trade-off is achieved when the least amount of energy is consumed in delivering the desired storage element performance.

As discussed earlier in this chapter, the clocking energy in a clocked storage element is the amount of clocking energy expended in clocking the internal nodes of the CSE. To evaluate the total clocking energy per clock cycle in the entire clock subsystem, one needs to add the energy consumed in the clock distribution network. The energy consumed in the clock distribution network depends on the total switched capacitance, which is determined by the total number of clocked storage elements on a chip and the input capacitance of their clock inputs, the total wiring capacitance, and the total switched capacitance of clock drivers, as given by Eq. (3.14):

$$C_{\text{distrib-net}} = N_{\text{FF}} \cdot C_{\text{in-}Clk,\text{FF}} + C_{\text{wire}} + C_{\text{sw-buff}} \tag{3.14}$$

The first term in Eq. (3.14) is constant for a given selection of storage elements. The last two terms depend on the buffer insertion/placement strategy, and should be minimized. The shorter the total wire length, the smaller the wiring capacitance, C_{wire}. If the wire lengths from clock drivers to clock sinks are not equal, there will be a clock skew. The absolute value of insertion delay from the root of the clock tree to the clock sinks is not so important, but it is very important that these delays are balanced within the clock-skew specification. This imposes a limit on how much extra wiring cost one has to incur in order to keep the clock skew within a given margin. In addition, there is an energy-performance trade-off between wide wires driving heavy nets and narrow wires with buffer repeaters. Therefore, the lower limit on clock distribution energy consumption per clock cycle is imposed by $C_{\text{distrib-net}}$ and by the targeted clock slope at the inputs of the storage elements.

3.3.2. Interface with Combinational Logic

As in driving the clock input of a storage element, one needs the parameters relevant for driving the storage element data input. The skew between the data inputs is not relevant as long as the data input signals arrive within setup/hold time specification. The parameters relevant to the combinational logic designer are therefore the CSE input data slope and input data capacitance. The data slope affects the performance and energy consumption of both the driving logic and the storage elements. Clock and data slopes are generally not equal.

CHAPTER 4

PIPELINING AND TIMING ANALYSIS

4.1. ANALYSIS OF A SYSTEM THAT USES A FLIP-FLOP

In order to properly analyze the timing parameters associated with the clocked storage elements, we need to analyze the timing situation in a pipelined system. We should start first with the simplest case of a flip-flop and the single clock used in the design. This situation is illustrated in Fig. 4.1. Much of the discussion presented here was taken from the paper by Unger and Tan (1986), with some minor changes in notation.

There are two events that we need to prevent:

1. *The data arrive too late to be captured reliably in the next cycle.* There are two possible scenarios here: either the data arrived far too late and are completely missed in the next cycle, or they are just sufficiently late to be violating the setup time requirement of the storage element, thus not assuring reliable capture.
2. *The data arrive too early (during the same cycle),* thus violating the hold time requirement for the flip-flip.

4.1.1. Late Data Arrival Analysis

We cannot assure that the data will be properly captured in either of the cases discussed in the last section, and therefore we are not able to guarantee reliable operation of the system. In order to perform a simple analysis of this system, let us assume that the clock skew and jitter together can cause the maximum

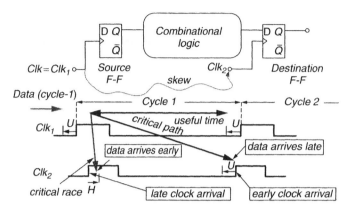

Figure 4.1. Timing in a digital system using a single clock and flip-flops.

deviation of the clock's leading edge for T_L amount of time from the nominal time of arrival (and T_T for the trailing edge). If we set the time reference to $t = 0$ for the leading edge of the clock for the Cycle 1, than we have a following relation for the latest data arrival:

$$t_{DLN} = T_L + D_{CQM} + t_{CR} \qquad (4.1)$$

In equations throughout this chapter, D_{CQm} represents the minimal clock-to-Q (output) delay of the flip-flop and D_{Lm} represents minimal delay through the logic (as opposed to the use of index M, where D_{CQM} and D_{LM} represent maximum delays).

The latest possible arrival of the data in the next cycle, t_{DLN}, occurs under the following circumstances: (1) data were captured at the latest possible moment due to the clock skew and jitter, which is T_L; (2) the flip-flip that captured the data was the slowest possible (keep in mind that flip-flop delays will vary due to the process variations); (3) these data traveled through the longest path in the logic, taking $t_L = t_{CR}$ (critical path):

$$t_{DLN} = P - T_L - U \qquad (4.2)$$

The clock's leading edge in the next cycle arrived at the earliest possible moment, $P - T_L$. However, in order to capture the data reliably, the data should arrive at least for the setup time, U, before the leading edge of the clock. This leads to the following inequality:

$$P - T_L - U \geq T_L + D_{CQM} + t_{CR} \qquad (4.3)$$

A constraint for the clock period, P (speed of the clock), is derived from this equality:

$$P \geq 2T_L + U + D_{CQM} + t_{CR} \qquad (4.4)$$

Alternatively for a given clock speed the longest critical path in the logic has to be shorter than

$$t_{CR} \leq P - 2T_L + U + D_{CQM} \tag{4.5}$$

This is one of the fundamental equations. Basically, it shows that the time available for information processing is equal to the time remaining in the clock period after the clock uncertainty is subtracted for both edges and the time data spent traveling through the storage element.

4.1.2. Early Data Arrival Analysis

A common misconception is that the flip-flop provides edge-to-edge timing and is thus easier to use, compared to the latch-based system, because it does not need to be checked for fast paths in the logic (*hold time violation*). This is not true, and the simple analysis that follows demonstrates that even with the flip-flop design, the fast paths can represent a hazard and invalidate the system operation.

If the clock controlling the flip-flop releasing the data is skewed so that it arrives early, and the clock controlling the flip-flop that receives these data arrives late, a hazard situation exists. This same hazard situation is present if the data travel through a fast path in the logic. A fast path is the path that contains very few to no logic blocks. Referring to Fig. 4.1, this hazard, which is also referred to as *critical race* (or race-through) can be described with the following set of equations:

$$t_{DEArr} = -T_L + D_{CQm} + D_{Lm} \tag{4.6}$$

$$t_{DEArrN} = +T_L + H \tag{4.7}$$

Equation (4.6) represents the time of the early arriving signal, t_{DEArr}, which should not be earlier than the time described by Eq. (4.7), otherwise there will be a hold-time violation of the data-receiving flip-flop. This condition is represented by the inequality (4.8):

$$-T_L + D_{CQm} + D_{Lm} > +T_L + H \tag{4.8}$$

Equation (4.8) gives us a limit on the fast paths, that is, no signal in the logic should be taking a time shorter than D_{LB}, otherwise, there will be hold-time violation in the circuit.

$$D_{Lm} > D_{LB} = 2T_L + H - D_{CQm} \tag{4.9}$$

Furthermore, the clock has to be active for some minimum amount of time (in order to assure reliable capture of data):

$$W \geq T_L + T_T + t_{CWm} \tag{4.10}$$

Equations (4.5), (4.9), and (4.10) provide timing requirements for the reliable operation of a system using flip-flops.

4.2. ANALYSIS OF A SYSTEM THAT USES A SINGLE LATCH

A system using a single latch is more difficult to analyze than a flip-flop-based system. This is because a single latch is transparent while the clock is active and the possibility for a race-through exists. However, this analysis is still much simpler than a general analysis of a system using two latches (MSL-based system), shown in Unger and Tan (1986). The use of a single latch represents a hazard due to the transparency of the latch, which introduces a possibility of races in the system. Therefore, the conditions for the single-latch-based system must account for critical race conditions. As the previous analysis shows, the presence of the storage element delay decreases the useful time in the pipeline cycle. Therefore, in spite of the hazards introduced by this design, the additional performance gain may well be worth the risk. This will be discussed in the following chapters.

Some well-known systems, such as the CRAY-1 supercomputer, use a single latch (Cray Research 1984). This decision was made for performance reasons. The second-generation Digital Corporation Alpha WD21164 processor uses single-latch-based design as well (Benschneider et al. 1995). One difference between Alpha and CRAY-1 is the way a single latch has been used in the pipeline. Two ways of structuring the pipeline with the single latch are shown in Fig. 4.2. Figure 4.2a shows a straightforward way of using a single latch. Here all the latches in the system are transparent while the clock is active (logic 1) and all the latches are *opaque* (nontransparent) when the clock is inactive (logic 0).

We will base the analysis of the single-latch-based design on the well-known paper by Unger and Tan (1986). Case (1) is easier to analyze, while case (2) becomes more complex. Case (2), also known as a *split-latch* design, will be explained by the example at the end of this chapter.

4.2.1. Late Data Arrival Analysis

In the case of a latch, the input signal needs to arrive at least a setup time, U, before the trailing edge of the clock (the edge that closes the latch). However, this edge could arrive earlier because of the clock skew. Therefore, the latest arrival of data that assures reliable capture after period P has to be

$$t_{DLArr} \leq W - T_T - U + P \tag{4.11}$$

Data captured at the end of the clock period could be the result of two events (whichever is the later):

1. The data were ready, clock arrived at the latest possible moment, T_L, and the worst-case delay of the latch, namely, D_{CQM}, was incurred.

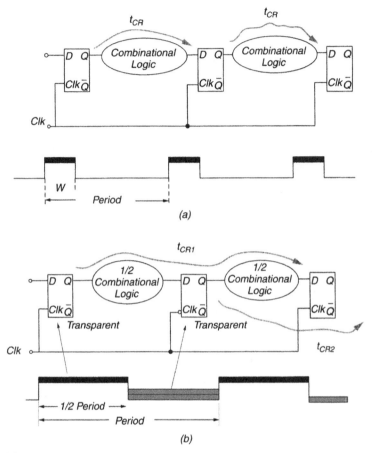

Figure 4.2. Two ways of using a latch in a single-latch-based system: (a) Case 1, (b) Case 2.

2. The clock was active and the data arrived at the last possible moment, which is a setup time, U, and clock skew time, T_T, before the trailing edge of the clock.

In both cases (1) and (2) the path through the logic was the longest path D_{LM}.
Thus in the worst scenario (either (case 1) or (case 2)) the data to be captured in the next cycle have to arrive in time to be reliably captured in the next cycle:

$$\max\{T_L + D_{CQM}, W - T_T - U + D_{DQM}\} + D_{LM} \geq W - T_T - U + P \quad (4.12)$$

This gives us a limit for the clock speed in terms of the duration of the period, P:

$$P \geq \max\{T_L + T_T + U + D_{CQM} - W, D_{DQM}\} + D_{LM} \quad (4.13)$$

This inequality breaks down into two inequalities, (4.14) and (4.15):

$$P \geq D_{LM} + D_{CQM} + T_L + T_T + U - W \qquad (4.14)$$

$$P_m = P \geq D_{LM} + D_{DQM} \qquad (4.15)$$

Equation (4.15) shows the minimal bound for P_m, which is the time it takes to traverse the loop, which consists of the maximum delay of the data passing through the latch and through the longest path in the logic. In other words: "Starting" from the leading edge of a clock pulse, there must be time, under worst case, before the trailing edge of the clock in the next cycle, for a signal to pass through the latch and the logic block in time to meet the setup time constraint" (Unger and Tan 1986).

The value of $P = P_m$ determines the highest frequency of the clock under which that particular system can operate reliably. However, this does not come without a price. Given that the loop through the logic and the latch is open, we have to be sure that any of the "fast paths" that may exist in the logic do not arrive sooner than the next period of the clock. This leads to the following analysis for fast paths.

4.2.2. Early Signal Arrival Analysis

The fastest signal traveling through the fastest path in the logic should arrive at least a hold time after the latest possible arrival of the same clock:

$$t_{DEArrN} \geq W + T_T + H \qquad (4.16)$$

There are two possible scenarios for the early arrival of the fast signal: (1) it was latched early and it passed through a fast path in the logic, or (2) it arrived early while the latch was transparent and passed through the fast latch and a fast path in the logic. This is expressed in Eq. (4.17):

$$t_{DEArrN} = \min\{t_{CEL} + D_{CQm}, t_{DEArr} + D_{DQm}\} + D_{Lm} \qquad (4.17)$$

The earliest arrival of the clock t_{CEL} happens when the leading edge of the clock is skewed to arrive early at $-T_L$. Thus, the condition for preventing race in the system is expressed as:

$$\min\{-T_L + D_{CQm}, t_{DEArr} + D_{DQm}\} + D_{Lm} \geq W + T_T + H \qquad (4.18)$$

The earliest possible arrival of the clock, plus clock-to-output delay of the latch, has to occur earlier than the early arrival of the data (while the latch is transparent), plus data-to-output delay of the latch. Thus, Eq. (4.18) becomes:

$$-T_L + D_{CQm} + D_{Lm} \geq W + T_T + H \qquad (4.19)$$

which gives us a lower bound on the permissible signal delay in the logic:

$$D_{Lm} > D_{LmB} \geq W + T_T + T_L + H - D_{CQm} \qquad (4.20)$$

Thus the conditions for the reliable operation of a system using a single latch are described by Eqs. (4.14), (4.15), and (4.20), which are repeated here for clarity:

$$P_m = P \geq D_{LM} + D_{CQM} + T_L + T_T + U - W \qquad (4.21)$$

$$P \geq D_{LM} + D_{DQM} \qquad (4.22)$$

$$D_{Lm} > D_{LmB} \geq W + T_T + T_L + H - D_{CQm} \qquad (4.23)$$

One can see from Eq. (4.21) that an increase in the clock width, W, can be beneficial for speed, but it increases the minimal bound for the fast paths, Eq. (4.23). The maximum useful value for W is obtained when the period P is minimum, Eq. (4.15). Substituting P from Eq. (4.22) into Eq. (4.21) yields the optimal value of W:

$$W_{opt} = T_L + T_T + U + D_{CQM} - D_{DQM} \qquad (4.24)$$

If we substitute the value of the optimal clock width, W_{opt}, into Eq. (4.21), then we will obtain the values for the maximum speed Eq. (4.25), and the minimum signal delay in the logic Eq. (4.26) that have to be maintained in order to satisfy the conditions for optimal single-latch system clocking:

$$P \geq D_{LM} + D_{DQM} \qquad (4.25)$$

$$D_{LmB} = 2(T_T + T_L) + H + U + D_{CQM} - D_{CQm} - D_{DQM} \qquad (4.26)$$

Equation (4.25) tells us that in a single-latch system, it is possible to make the clock period, P, as small as the sum of the delays in the signal path: latch and critical path delay in the logic block. This can be achieved by adjusting the clock width, W, and assuring that all the fast paths in the logic are larger in their duration than some minimal time, D_{LmB}. In practice, the optimal clock width, W_{opt} is very small and can support the use of pulsed-latches.

It might be worthwhile thinking about the meaning of Eq. (4.25) and (4.26). What Eq. (4.26) tells us is that under ideal conditions, if there are no clock skews and no process variations, the fastest path through the logic has to be greater than the sampling window of the latch $(H + U)$ minus the time the signal spends traveling through the latch. If the travel time through the latch, D_{DQM} is equal to the sampling window, than we do not have to worry about fast paths. This is the case of the race immunity, $R \geq 0$. Of course, in practice, we do have to take care of both fast and slow paths in the logic.

4.3. ANALYSIS OF A SYSTEM WITH A TWO-PHASE CLOCK AND TWO LATCHES IN AN M–S ARRANGEMENT

A particular version of the use of two latches in the M–S configuration is the most commonly used technique in digital system design. It is also a robust and reliable technique compatible with the design for testability (DFT) methodology. We will start by describing the most general arrangement, consisting of two latches clocked by two separate and independent clocks ϕ_1 and ϕ_2, as shown in Fig. 4.3.

Analysis of a system using a two-phase clock is much more complex compared to the system using a single clock, because we are introducing skew on the second clock. Therefore the set of parameters includes clock skew on both the leading and trailing edges of the first clock ϕ_1, T_{1L} and T_{1T}, and on the second clock ϕ_2, T_{2L} and T_{2T}. In addition, the overlap, V, between ϕ_1 and ϕ_2 is to be taken into account as are the corresponding widths of the clock pulses, W_1 and W_2.

This analysis tends to be tedious and complex. It is therefore suggested that the interested reader give the paper by Unger and Tan (1986) a detailed reading. Without going into the details of that analysis, we present only a qualitative analysis and final derivations.

Several conditions can be derived from the latest signal arrival analysis. First, we need make sure there is an orderly transfer into latch L_2 (slave) from latch

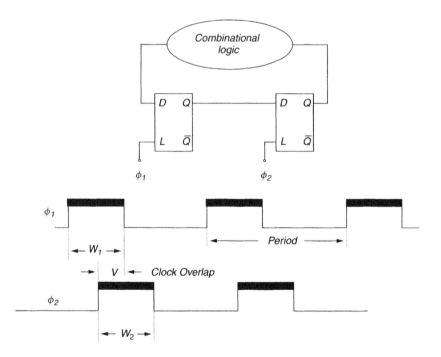

Figure 4.3. System using two-phase clock and two latches in M–S arrangement.

L_1 (master), even if the signal arrives late (at the last possible moment) in the latch L_1 (master). This analysis yields the following two conditions:

$$W_2 \geq V + U_2 - U_1 + D_{1DQM} + T_{2T} + T_{1L} \tag{4.27}$$

$$W_1 + W_2 \geq V + U_2 + D_{1CQM} + T_{1T} + T_{2T} \tag{4.28}$$

These conditions assure the timely arrival of the signal in the latch L_2, and thus an orderly L_1–L_2 transfer (from master to slave).

The analysis of the latest arrival of the signal into the latch L_1 in the next cycle (critical path analysis) yields Eq. (4.29), (4.30), and (4.31):

$$P \geq D_{1DQM} + D_{2DQM} + D_{LM} \tag{4.29}$$

Equation (4.29) gives us the highest frequency at which the system can operate. In other words, the minimum period of clock P has to be of sufficient duration to allow the signal to traverse the loop consisting of latch L_1, latch L_2, and the longest path in the logic, D_{LM}:

$$W_1 \geq -P + D_{1CQM} + D_{2DQM} + U_1 + D_{LM} + T_{1L} + T_{1T} \tag{4.30}$$

The condition specified in Eq. (4.30) assures the timely arrival of the signal that starts on the leading edge of ϕ_1, traverses the path through L_2, which is the longest path in the logic, and arrives to L_1 before the trailing edge of ϕ_1, in time to be captured.

If the signal, starting from the leading edge of ϕ_2 (prior to the end of ϕ_1) traversing L_2 and the longest path in the logic, is to be captured in time in L_1, then the condition Eq. (4.31) needs to be satisfied.

$$P \geq -V + D_{2CQM} + U_1 + D_{LM} + T_{1T} + T_{2L} \tag{4.31}$$

Equation (4.31) shows that the amount of overlap, V, between clocks ϕ_1 and ϕ_2 has some positive effect on speed. The overlap, V, allows the system to run at greater speed. Conversely, if we increase V, we can tolerate a longer critical path, D_{LM}. Thus, the increase in V is beneficial to the system. However, the increase in the clock overlap has its negative effects and its limitations. One of the negative consequences is that overlapping clocks introduce the possibility of race conditions, thus requiring a fast-path analysis. The analysis of fast paths (or critical races) makes the timing analysis much more complex, and CAD tools generally do not perform this analysis very well. For that reason one would sacrifice some performance for reliability and ease of design. In a robust design that avoids fast paths, nonoverlapping clocks are used. Those nonoverlapping phases of the clock are usually generated locally, to avoid the difficulty in distributing two phases of the clock throughout the system. One commonly used clocking

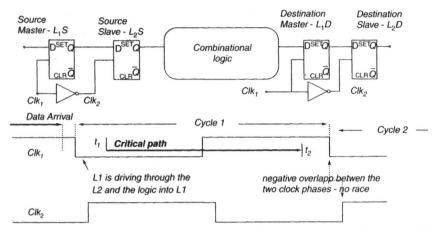

Figure 4.4. M–S $(L_1-L_2$ latch) with nonoverlapping clocks Φ_1 and Φ_2 obtained by locally generating clock Φ_2. (This arrangement is also commonly referred to as flip-flop.)

methodology is to use MSLs (L_1-L_2) with the locally generated ϕ_2 clock. Such an arrangement, show in Fig. 4.4, assures reliability, since the ϕ_1 and ϕ_2 clocks are not overlapped. Thus, for the vast majority of practical cases it eliminates the need for critical race analysis. The apparent flip-flop-like behavior of this configuration has caused the term "flip-flop" to be widely used, although the structure is actually an MSL (L_1-L_2).

High-performance systems are designed with the objective of maximizing performance. Therefore, clocks ϕ_1 and ϕ_2 are commonly overlapped, thus leading to the critical-race analysis (again, the reader is referred to the Unger and Tan paper). The analysis suggests limiting the minimum signal delay in logic D_{LmB} in order to prevent the critical race:

$$D_{Lm} > D_{LmB} = V + H_1 + T_{1T} + T_{2L} - D_{2CQm} \qquad (4.32)$$

Equation (4.32) tells us that any amount of time we have added to the upper bound of the critical path, giving us more time in the logic, will have to be added to the minimal bound for the short paths, which increases the limit on the short path. This may force us to add some padding to the short paths (insert inverters in order to increase the delay) in order to meet the constraint (4.32).

It is interesting to know the maximal amount of overlap, V, that can be used. This is obtained by solving the timing equations Eqs. (4.29) and (4.31) (Unger and Tan 1986), leading to Eq. (4.33):

$$V_{\max} = T_{1T} + T_{2L} + D_{2CQM} + U_1 - D_{1DQM} - D_{2DQM} \qquad (4.33)$$

In summary, when using a two-phase clock with MSLs (L_1-L_2), a conservative design would eliminate the need for analysis of the fast paths (critical race condition). This design is arrived at by using nonoverlapping clocks ϕ_1 and ϕ_2.

However, this reliability is achieved at the expense of performance. When maximum performance is the objective, it is possible to adjust the clock overlap, V, by phasing clocks ϕ_1 and ϕ_2 so that the system runs at the maximum frequency. Maximum clock frequency is reached when P_{min} is equal to the sum of the delays incurred when traversing the path consisting of the maximum logic delay and delays in latches L_1 and L_2.

Example: Analysis of the First-Generation Alpha Processor (WD21064)

An appropriate example of the optimal clock parameters of a system using a single latch is the first-generation Alpha processor, Fig. 4.5a and 4.5b.

A description of the system is presented in the paper by Dobberpuhl et al. (1992). We will use notation adopted from Unger and Tan (1986) and assume the following system parameters for the sake of an example:

Clock skew: $T_L = T_T = 20$ ps, for both edges of the clock.

Latch L_1 parameters are clock-to-Q delay $D_{CQM} = 50$ ps; $D_{CQm} = 30$ ps; D-to-Q delay $D_{DQM} = 60$ ps; setup time $U = 20$ ps; hold time $H = 30$ ps.

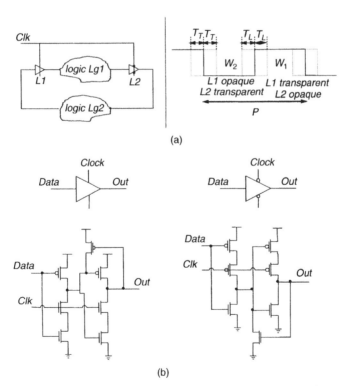

(a)

(b)

Figure 4.5. (a) Timing arrangement used in the first-generation Alpha processor. (b) Latches used in the first-generation Alpha processor. (Dobberpuhl et al. 1992), Copyright © 1992 IEEE.

Latch L_2 parameters are: $D_{CQM} = 60$ ps; $D_{CQm} = 40$ ps; $D_{DQM} = 70$ ps; $U = 30$ ps; $H = 40$ ps.

The structures of the L_1 and L_2 latches used in the second generation of the Alpha processor are shown in Fig. 4.5b.

The critical paths in logic Sections 1 and 2 are $D_{L1M} = 200$ ps and $D_{L2M} = 170$ ps.

For the given clock setup, $V = 0$ and, clearly, $P = W_1 + W_2$.

With the nominal time, $t = 0$, set at the leading edge of the clock, we obtain the latest allowed data arrival times into latches L_1 and L_2, respectively:

$$t_{D1LArr} \leq W_1 - T_T - U_1 \tag{4.34}$$

$$t_{D2LArr} \leq P - T_L - U_2 \tag{4.35}$$

The latest arrival time of the data into latch L_2 is limited by the time at which latch L_1 releases the data into the logic stage Logic$_1$:

$$t_{D2LArr} = \max\{t_{D1LArr} + D_{1DQM}, T_L + D_{1CQM}\} + D_{L1M} \tag{4.36}$$

Combining Eq. (4.34), Eq. (4.35), and Eq. (4.36), we obtain

$$W_1 - T_T - U_1 + D_{1DQM} + D_{L1M} \leq P - T_L - U_2 \tag{4.37a}$$

$$T_L + D_{1CQM} + D_{L1M} \leq P - T_L - U_2 \tag{4.37b}$$

Rearranging Eqs. (4.37a) and (4.37b), we obtain a set of bounds for W_2 and P:

$$W_2 \geq U_2 - U_1 + D_{1DQM} + T_L - T_T + D_{L1M} \tag{4.38a}$$

$$P \geq U_2 + D_{1CQM} + 2T_L + D_{L1M} \tag{4.38b}$$

Because of the symmetry of the clocking scheme, moving the reference point from the clock's leading edge to its trailing edge will give us the same equations with indexes interchanged. To check this, start from the equation analogous to Eq. (4.36):

$$t_{D1LArr} = \max\{t_{D2LArr} + D_{2DQM}, -W_2 + T_T + D_{2CQM}\} + D_{L2M} \tag{4.39}$$

Combining Eq. (4.34) and Eq. (4.39) and rearranging, we obtain a set of bounds for W_1 and P:

$$W_1 \geq U_1 - U_2 + D_{2DQM} + T_T - T_L + D_{L2M} \tag{4.40a}$$

$$P \geq U_1 + D_{2CQM} + 2T_T + D_{L2M} \tag{4.40b}$$

Combining Eq. (4.38a) and Eq. (4.40a) we obtain a third and often the most critical bound for the clock period:

$$P = W_1 + W_2 \geq D_{1DQM} + D_{L1M} + D_{2DQM} + D_{L2M} \tag{4.41}$$

Substituting the values into Eqs. (4.38), (4.40), and (4.41), we obtain:

$$W_2 \geq 270 \text{ ps}$$

$$P \geq 320 \text{ ps}$$

$$W_1 \geq 230 \text{ ps}$$

$$P \geq 270 \text{ ps}$$

and the most critical bound for P,

$$P = W_1 + W_2 \geq 500 \text{ ps}$$

Thus the minimum clock period is $P_{\min} = 500$ ps, and the maximum frequency at which this system can run is $f_{\max} = 2$ GHz.

4.4. ANALYSIS OF A SYSTEM WITH A SINGLE-PHASE CLOCK AND DUAL-EDGE-TRIGGERED STORAGE ELEMENTS

A dual-edge-triggered storage element (DETSE) is so named because it captures its data at both clock edges. The timing parameters of this storage element are defined for both clock edges, and have the same meaning as those for the single-edge-triggered storage element.

Since DETSE is an edge-sensitive storage element, the analysis of dual-edge triggered system with a single-phase clock is similar to that described in Section 4.1. A diagram of a system using DETSE is shown in Fig. 4.6. The single-phase clock is specified by its period, P; duty cycle (clock pulse width relative to period), w; clock pulse width, W, which is equal to w^*P; and maximum clock uncertainty for leading and trailing clock edges, T_L and T_T, respectively. For each storage element, $D_{CQM,L}$, $D_{CQm,L}$, U_L, H_L, $D_{CQM,T}$, $D_{CQm,T}$, U_T, and H_T, designate maximum and minimum clock-to-output delay, setup time, and hold time, where indices L and T stand for leading and trailing edge of the clock, respectively. Nonclocked logic blocks between storage elements have maximum and minimum delays, D_{LM} and D_{Lm}, respectively. Note that when CSE_1 releases data at the leading edge of the clock, CSE_2 captures it at the trailing edge, and vice versa. Therefore there are two scenarios for each of the two clock edges that have to be prevented.

1. Data reaches the destination storage element too late to be captured. This scenario is prevented if the data are scheduled to arrive at the destination storage element at the latest at a setup time prior to the capturing clock edge.

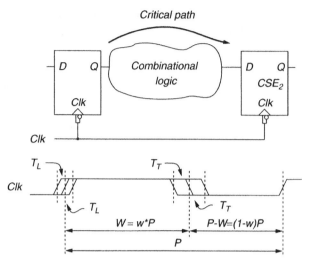

Figure 4.6. Digital system using a single-phase clock and dual-edge triggered storage elements.

2. Data reaches the destination storage element early enough to corrupt the safe capture with the same edge that released the data. This scenario is prevented by ensuring that data arrives to the destination storage element a hold time after the clock edge.

The setup and hold time requirements for the two clock edges provide four basic conditions that assure safe operation in dual-edge clocking systems (Nedovic and Oklobdzija 2001). We will examine each of these conditions closely.

4.4.1. Late Data Arrival

First, data from source CSE at the leading edge of the clock must arrive to destination CSE early enough to be safely captured by the trailing edge of the clock. This requirement must be met even with the worst storage element and logic delay, and clock uncertainty. Safe capture of the data occurs only if the data arrive at the input of the destination CSE at the latest setup time before the trailing edge of the clock. From Fig. 4.6, we see that the arrival of the trailing edge of the clock is delayed from the releasing clock edge for as long as the clock is at its high level:

$$T_L + D_{CQM,L} + D_{LM} + T_T + U_T \leq wP \qquad (4.42)$$

Similarly, if the data are released at the falling edge and captured at the rising edge of the clock, a similar relation holds:

$$T_T + D_{CQM,T} + D_{LM} + T_L + U_L \leq (1 - w)P \qquad (4.43)$$

Equations (4.42) and (4.43) determine the minimum clock period for the given duty cycle:

$$
P \geq P_m = \max
$$
$$
\times \left(\frac{T_L + D_{CQM,L} + D_{LM} + U_T + T_T}{w}, \frac{T_T + D_{CQM,T} + D_{LM} + U_L + T_L}{1 - w} \right)
$$

$$(4.44)$$

Alternatively, if the clock period, P, is specified, the preceding analysis provides the maximum allowable logic delay:

$$
D_{LM} \leq \min(wP - (D_{CQM,L} + U_T), (1 - w)P
$$
$$
-(D_{CQM,T} + U_L)) - T_L - T_T
$$

$$(4.45)$$

It is not always possible to control clock duty cycle. Therefore an important special case for all practical purposes is the symmetric clock ($w = 0.5$). Use of the symmetric clock simplifies clock generation and reduces clock uncertainties. In the case where $w = 0.5$, Eqs. (4.44) and (4.45) become:

$$
P \geq P_m = 2(\max(D_{CQM,L} + U_T, D_{CQM,T} + U_L) + D_{LM} + T_L + T_T) \quad (4.46)
$$

$$
D_{LM} \leq P/2 - \max(D_{CQM,L} + U_T, D_{CQM,T} + U_L) - T_L - T_T \quad (4.47)
$$

In the general case, an optimum point can be found by using clock duty cycle w, which minimizes P and satisfies both Eq. (4.42) and Eq. (4.43):

$$
w_{opt} = \frac{D_{CQM,L} + U_T + D_{LM} + T_L + T_T}{D_{CQM,L} + D_{CQM,T} + U_L + U_T + 2D_{LM} + 2T_L + 2T_T} \quad (4.48)
$$

The corresponding minimum achievable clock period is:

$$
P_m = D_{CQM,L} + D_{CQM,T} + U_L + U_T + 2D_{LM} + 2T_L + 2T_T \quad (4.49)
$$

Again, if the clock period, P, is specified and the goal is to find maximum logic delay, Eqs. (4.48) and (4.49) become Eqs. (4.50) and (4.51):

$$
w_{opt} = \frac{1}{2} \left(\frac{D_{CQM,L} - D_{CQM,T} + U_T - U_L}{P} + 1 \right) \quad (4.50)
$$

The corresponding maximum logic delay is:

$$
D_{LM.\max} = \frac{1}{2}(P - (D_{CQM,L} + D_{CQM,T} + U_L + U_T + 2T_L + 2T_T)) \quad (4.51)
$$

This analysis shows that in the general case the clock needs to be asymmetric ($w \neq 0.5$) for optimum operation. The measure of this asymmetry is the difference in the costs of DETSE in two half-periods of the clock, Eqs. (4.48) and (4.50). Since the optimum duty cycle does not depend on the logic delay, it is the same for all the paths in the system.

Equations (4.48) and (4.50) also indicate how to achieve optimum operation with the arbitrary clock duty cycle. As mentioned earlier, we are mainly interested in the symmetric clock ($w = 0.5$). From Eq. (4.48), the condition for $w_{opt} = 0.5$ leads to:

$$D_{CQM,L} + U_T = D_{CQM,T} + U_L \tag{4.52}$$

Thus, a requirement for a good design of DETSE with symmetric clock is to closely comply with Eq. (4.52). In order to obtain most performance out of DETSE, it is necessary to minimize both sides of Eq. (4.52), as both present the timing overhead of the storage element.

4.4.2. Early Data Arrival

In addition to the preceding, it must be certain that the data arrive at the destination CSE late enough to prevent its hold-time violation. Correct operation must be set even for the earliest allowed arrival of the releasing clock edge, which is minimum clock-to-output and logic delay, and the latest arrival of the capturing clock edge. Since releasing and capturing clock edges occur simultaneously, this failure mechanism is the same as with the single-edge-triggered storage elements described in Section 4.1. For the leading edge of the clock:

$$-T_L + D_{CQm,L} + D_{Lm} \geq T_L + H_L \tag{4.53}$$

Similarly, for the following relation has to be satisfied for the trailing edge of the clock:

$$-T_T + D_{CQm,T} + D_{Lm} \geq T_T + H_T \tag{4.54}$$

Equations (4.53) and (4.54) determine the minimum logic delay in the stage to avoid a hold-time violation:

$$\boxed{D_{Lm} \geq \max((H_L + 2T_L - D_{CQm,L}), (H_T + 2T_T - D_{CQm,T}))} \tag{4.55}$$

Thus, for a given clock period and duty cycle, Eqs. (4.45) and (4.55) provide a set of requirements for reliable operation.

Example The following example illustrates the use of the dual-edge clocking strategy in a pipelined system and shows how to maximize performance. For the two-stage dual-edge-triggered system shown in Fig. 4.7, the timing parameters of the storage elements used are clock-to-output delay $D_{CQM,L} = 150$ ps, $D_{CQm,L} = 80$ ps, $D_{CQM,T} = 200$ ps, $D_{CQm,T} = 150$ ps, $U_L = 50$ ps, $U_T = 0$, $H_L = 100$ ps, $H_T = 40$ ps. The clock uncertainty is $T_L = 20$ ps for the leading

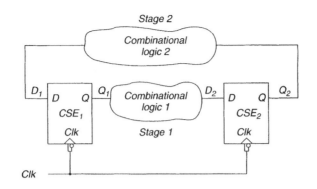

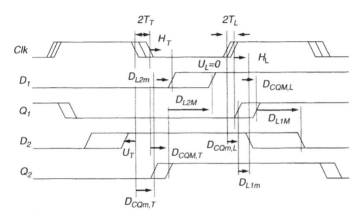

Figure 4.7. Two-stage dual-edge-triggered system.

edge of the clock, and $T_T = 40$ ps for trailing edge of the clock. The delays of the critical paths in the two logic blocks are $D_{LM1} = 900$ ps and $D_{LM2} = 950$ ps, respectively. The goal is to find the minimum logic delays and minimum clock period that still allow safe operation, both for duty cycle $w = 50\%$ and for the optimum duty cycle.

In order to meet the hold-time requirement for storage element CSE_1, Eq. (4.55) must be satisfied:

$$D_{Lm1}, D_{Lm2} \geq \max(100 + 20 + 20 - 80, 40 + 40 + 40 - 150) = 60 \text{ ps} \quad (4.56)$$

If the duty cycle is 50%, the minimum clock period that allows safe operation must satisfy Eq. (4.46). For Stage 1:

$$P \geq 2(\max(150 + 0, 200 + 50) + 900 + 20 + 40)$$

$$= \max(2220, 2420) = 2420 \text{ ps} \quad (4.57)$$

For Stage 2:

$$P \geq 2(\max(150 + 0, 200 + 50) + 950 + 20 + 40)$$
$$= \max(2320, 2520) = 2520 \text{ ps} \qquad (4.58)$$

The minimum clock period is imposed by the setup-time requirement for the leading edge of the clock at CSE_1. With this clock period, the latest data arrival with respect to the trailing edge of the clock is still 100 ps prior to setup time. Thus, CSE_1 is not optimal to use with the symmetric clock, since a lag of 100 ps exists that is used by neither the logic nor the storage element. One way to use this lag and further reduce the clock period is to reduce the duration of the high level of the clock and to keep the duration of the low level of the clock the same. According to Eqs. (4.48) and (4.49), the optimum duty cycle and minimum clock period achieved in this way are

$$w_{opt} = \frac{150 + 0 + 950 + 20 + 40}{150 + 200 + 50 + 0 + 2 \cdot 950 + 2 \cdot 20 + 2 \cdot 40}$$
$$= \frac{1160}{2420} = 47.9\% \qquad (4.59)$$
$$P_m = 150 + 200 + 50 + 0 + 2 \cdot 950 + 2 \cdot 20 + 2 \cdot 40 = 2420 \text{ ps} \quad (4.60)$$

Thus, tuning the duty cycle is a way of neutralizing the imbalance in the storage element timing parameters for the leading and trailing edge of the clock. Figure 4.8 shows the achievable clock periods versus duty cycle for the example from Fig. 4.7. If the duty cycle and the clock period are in the allowed region, the setup-time requirement for both clock edges is met. For the low duty cycle, the clock period must be increased in order to meet the setup requirement for the trailing edge of the clock. Similarly, if the duty cycle is higher than optimal,

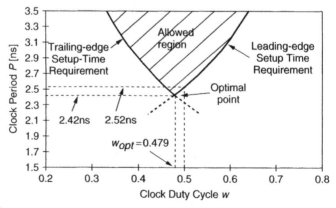

Figure 4.8. Allowed clock period as a function of the clock duty cycle in the dual-edge-triggered system of Fig. 4.7.

the clock period must be increased so that the setup requirement for the leading edge of the clock is met.

Another, more practical way to increase the allowable time in the logic is to keep the clock symmetric and redesign the storage elements so that Eq. (4.52) is satisfied. This can be achieved by transistor resizing or changes in the topology of the DETSE. In this example, if it is possible to reduce $D_{CQM,T}$ to 150 ps at the expense of increasing $D_{CQM,L}$ to 200 ps, the optimum duty cycle would be:

$$w_{opt} = \frac{200 + 0 + 950 + 20 + 40}{200 + 150 + 50 + 0 + 2 \cdot 950 + 2 \cdot 20 + 2 \cdot 40} = 0.5 \qquad (4.61)$$

The minimum achievable clock period is the same as that achieved by tuning of the duty cycle:

$$P \geq \max \left(\frac{200 + 0 + 950 + 20 + 40}{0.5}, \frac{150 + 50 + 950 + 20 + 40}{0.5} \right) = 2420 \text{ ps} \qquad (4.62)$$

Thus, both tuning the duty cycle and optimal design of the storage elements allow the clock period to be minimized. In this example, the system can run at the maximum frequency of $f = 413$ MHz.

CHAPTER 5

HIGH-PERFORMANCE SYSTEM ISSUES

Clocking in high-performance digital systems is most seriously affected by *clock skew* and *clock jitter*. In the past, clock skew was the dominant factor. Recently, however, clock jitter has started gaining dominance over clock skew. Here we will treat both of them as *clock uncertainties*. With the recent trend in frequency scaling, the number of logic gates per stage decreases and the pipeline becomes deeper, so that the portion of the clock cycle budgeted for clock uncertainty increases. In addition, production and distribution of the high-frequency clock to the increasing number of storage elements becomes progressively difficult due to various issues, such as load mismatch, power supply and substrate noise, and temperature variations. As a result, clock uncertainties occupy an increasing portion of the cycle time. The ability to reduce the impact of these uncertainties is one of the most important properties of the high-performance system.

The second important issue in high-performance digital systems is variation of the signal delays and the ability to absorb the delay of a signal that stretches beyond the time allotted to it by the pipeline stage. The ability of the pipeline to be flexible, thus allowing the extra delay to be absorbed by subsequent pipeline stages, without disrupting the correct operation is essential.

5.1. ABSORBING CLOCK UNCERTAINTIES

The clock uncertainties were of little consequence in the 1970s and 1980s, but in modern designs they are a limitation to further performance scaling (Heald et al. 2000a; Hofstee et al. 2000; Harris and Horowitz 1997). As an illustration, the time budget allotted to clock uncertainties is typically on the order of one to two FO4 inverter delays in modern microprocessors. This usually accounts for more than 10% of the entire clock cycle.

It is common practice in the VLSI circuit design to consider the clock uncertainties as an inevitable timing cost. In this approach, the only way to reduce the impact of the clock uncertainties is to either reduce the clock speed or to reduce the uncertainties themselves. There are several methods that can be used to minimize the jitter or skew components of clock uncertainties. Typically, jitter is minimized using better clock generators, or by reducing the noise in the power supply of the clock buffers, while skew is reduced with careful clock distribution or active deskewing. Alternatively, the impact of clock skew is minimized when all the critical paths are placed in the same clock domain. However, these techniques become increasingly difficult because of poor scaling of the clock uncertainty.

The useful time available for computation within each clock cycle is nominally reduced by the CSE overhead. This overhead changes as a function of clock uncertainty. The change is smaller than the clock uncertainty itself. This decrease in uncertainty defines an important property named *clock uncertainty absorption*.

A recent flip-flop design, controlled by a narrow, locally generated clock pulse, with negative setup time, exhibits some degree of clock uncertainty absorption (Partovi et al. 1996). This is shown by the relationship between the clock *Clk* and the output *Q*, in the presence of clock jitter, as illustrated in Fig. 5.1. The variation in the arrival time of the clock is somewhat absorbed by the flip-flop, resulting in a *smaller* variation in the time at which the output changes. This behavior can be explained as follows. If the capturing pulse is sufficiently wide, the flip-flop is briefly transparent to the data signal, resembling latch behavior, and its timing is

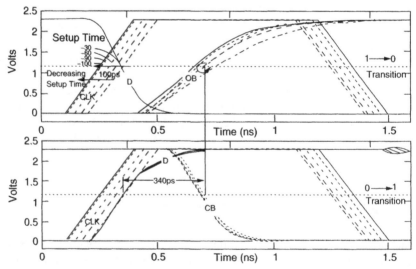

Figure 5.1. Output of a flip-flop in the presence of clock jitter. (Partovi et al. 1996.), Copyright © 1996 IEEE.

less sensitive to the clock arrival. This short transparency period is also known as the *soft clock edge*. With the increased importance of clock uncertainties, the practical use of the clocked storage element in high-performance systems will depend, to a large extent, on its ability to absorb them.

Typically, clock uncertainty absorption can be achieved with a level-sensitive clocking strategy. By definition, an edge-sensitive clocking strategy is based on CSEs triggered with a fixed ("hard") clock edge, and is not suitable for clock uncertainty absorption. However, an edge-sensitive clocking strategy using flip-flops with a soft clock-edge property allows a certain level of clock uncertainty absorption. In the rest of this section we discuss the clock-uncertainty absorption of the soft clock-edge flip-flops. The clock absorbing properties of level-sensitive clocking using transparent latches is addressed in Section 5.3.

5.1.1. Clock-Uncertainty Absorption Using Soft Clock Edge

The clock uncertainties are manifested as a variation in the arrival time of the clock edge. Typically, clock uncertainties are illustrated by using a time window that captures the occurrence of the clock edge. A clocked storage element absorbs uncertainties when the time of the output transition is not significantly affected by the variations in the arrival of the triggering edge of the clock.

To understand the effect of clock uncertainties, one should analyze the delay characteristic of the clocked storage element defined in Chapter 3. This characteristic represents data-to-output delay as a function of clock arrival time for a fixed data arrival time, as shown in Fig. 5.2. By observing how the output changes when the clock uncertainties are present, we are able to see how the uncertainties affect the delay of the storage element.

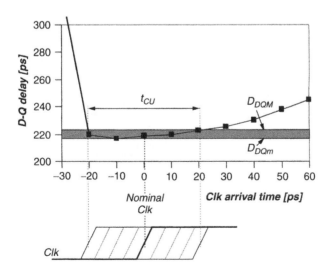

Figure 5.2. Data-to-output characteristics in the presence of clock uncertainty.

Figure 5.2 shows data-to-output delay versus clock arrival time when the data arrival time is constant. When no clock uncertainties are present, the clock is scheduled to arrive so that $D-Q$ delay (t_{DQm}) is smallest, in order to minimize the cost introduced by the clocked storage element.

Any variation in the clock arrival time increases the $D-Q$ delay. Given some maximum allowed $D-Q$ delay (D_{DQM}), we can use the delay characteristic of the CSE to find the corresponding clock uncertainty window, as shown in Fig. 5.2. The points at which D_{DQM} intersects with the delay characteristic determine the earliest and latest clock arrival times that are allowed. The ratio of the maximum variation of $D-Q$ delay ($D_{DQM} - D_{DQm}$) and width of the allowed clock uncertainty window t_{CU} illustrates the clock absorption property of the CSE. Since clock uncertainties are typically symmetric, we can also find the new optimum clock arrival time as the mean of the earliest and latest allowed clock arrival times. Using this methodology, the clock uncertainties are incorporated in the delay of the CSE. The $D-Q$ delay can be expressed as $D_{DQ} = D_{DQ}(t_{CU})$, and clock-to-output delay as $D_{CQ} = D_{CQ}(t_{CU})$, where t_{CU} is the clock uncertainty.

The key role of a CSE is to minimize the propagation of clock uncertainty to the CSE output. This can be characterized by the marginal increase in the $D-Q$ delay with respect to the amount of clock uncertainty. For data arriving at the nominal time, we find the worst-case $D-Q$ delay when the clock is allowed to arrive anywhere in the uncertainty window. This is illustrated in Fig. 5.3 where the worst-case delay is the maximum of the $D-Q$ delays over all clock arrivals. As in Fig. 5.2, the earliest and latest clock arrivals within the clock uncertainty window cause the latest change in the output Q of the CSE, as shown in Fig. 5.3. The maximum $D-Q$ delay, D_{DQM}, is defined at some optimal

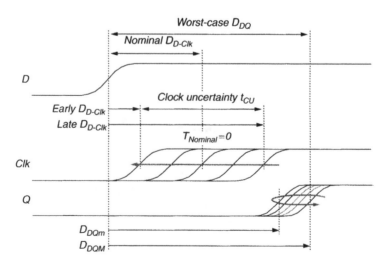

Figure 5.3. Dependence of data-to-output delay on clock arrival.

data arrival with respect to the nominal clock. This optimal data arrival, or optimal setup time, U_{opt}, is the data-to-nominal-clock delay that yields the smallest worst-case $D-Q$ delay over all clock arrivals in the uncertainty window, that is, minimizes Eq. (5.1) (Saint-Laurent et al. 2002):

$$\max_t[D_{DQ}(D_{D\text{-}CLK} + t)], \qquad t \in [-t_{CU}/2, t_{CU}/2] \qquad (5.1)$$

Consequently,

$$D_{DQM} = \max_t[D_{DQ}(U_{opt} + t)], \qquad t \in [-t_{CU}/2, t_{CU}/2] \qquad (5.2)$$

Note that the setup time and minimum $D-Q$ delay, D_{DQm}, as defined in Section 3.1.2, are special cases of Eqs. (5.1) and (5.2) when $t_{CU} = 0$. The increase in $D-Q$ delay due to the presence of the clock uncertainties is generally smaller than the amount of the uncertainty itself (Figs. 5.2 and 5.3). We express *clock-uncertainty absorption*, α_{CU}, of a storage element as the portion of the total clock uncertainty not reflected at the output:

$$\alpha_{CU} = \frac{t_{CU} - (D_{DQM} - D_{DQm})}{t_{CU}} = 1 - \frac{\Delta D_{DQ}}{t_{CU}} \qquad (5.3)$$

The relationship between D_{DQM} and D_{DQm}, and thus α_{CU}, is determined by t_{CU} and the $D-Q$ characteristic, $D_{DQ}(D_{D\text{-}CLK})$. As shown in Fig. 5.4, both clock uncertainty absorption and optimal setup time are largely dependent on the clock uncertainty. For small values of the clock uncertainty, it is possible to set the data arrival so that $D-Q$ delay does not change significantly regardless of the clock arrival. Equivalently, clock uncertainty absorption is high and optimal setup time is small, as illustrated in Fig. 5.4a. As the clock uncertainty increases, the $D-Q$ delay increases slowly, as long as the clock arrives within the relatively flat region around D_{DQm}, as shown in Fig. 5.2. As the clock arrives outside this region, D_{DQ}

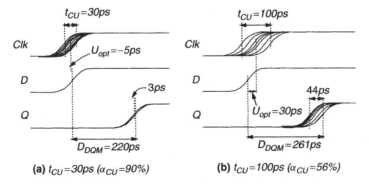

(a) $t_{CU}=30ps$ ($\alpha_{CU}=90\%$)   **(b)** $t_{CU}=100ps$ ($\alpha_{CU}=56\%$)

Figure 5.4. Total delay versus clock uncertainty.

increases more rapidly. Eventually, the clock arrivals enter the region where the delay characteristic has a slope of unity, that is, where the clock-to-output delay is constant. In this region, any uncertainty of the clock arrival directly propagates to the output Q. As a result, clock-uncertainty absorption, α_{CU}, decreases and the optimal setup time increases, as shown in Fig. 5.4b. The clock-uncertainty absorption is the most effective in the case where the clock arrivals can be contained in the window where the $D-Q$ characteristic is relatively flat.

5.1.2. Timing Analysis with Clock-Uncertainty Absorption

Late Data Arrival   Clocked storage elements that have the clock-uncertainty absorption property need to have a transparency window of a certain width. In level sensitive systems, this window is slightly smaller than half the clock period, while in edge-triggered systems with the soft-edge property, the transparency window is much shorter. Timing analysis of any system with a transparency window has to include data arrivals from multiple pipeline stages. Detailed timing analysis of a level sensitive system will be given in Sections 5.2 and 5.3.

In order to illustrate the impact of the clock-uncertainty absorption to the cycle time, we consider a simple case where equal clock uncertainty, t_{CU}, applies to identical flip-flops in all pipeline stages. Because the flip-flops have a soft-edge property, the timing analysis can be performed in a similar way to the single latch-based system in Chapter 4. Using the result of that analysis, Eq. (4.15), the following holds:

$$D_{DQM} + D_{LM} \leqslant P \tag{5.4}$$

The maximum $D-Q$ delay, D_{DQM}, is determined by Eq. (5.3) in terms of the effect of the clock uncertainty, t_{CU}. Expressing D_{DQM} from Eq. (5.3) yields

$$D_{DQM} = D_{DQ}(t_{CU}) = D_{DQm} + (1 - \alpha_{CU})t_{CU} \tag{5.5}$$

where $D_{DQ}(t_{CU})$ is $D-Q$ delay when the clock uncertainty, t_{CU}, exists, and D_{DQm} is the minimum $D-Q$ delay when there is no clock uncertainty. Assuming that the worst-case logic delay, D_{LM}, is the same in every stage, combining Eq. (5.4) and Eq. (5.5), we obtain the requirement for the minimum clock period in a system that uses flip-flops with soft clock edge:

$$D_{DQm} + (1 - \alpha_{CU})t_{CU} + D_{LM} \leqslant P \tag{5.6}$$

By comparing Eq. (5.6) to the case without the uncertainties, it can be seen that the only difference is in the factor $(1 - \alpha_{CU})t_{CU}$. Therefore, the total impact of the clock uncertainty on the clock cycle time is $(1 - \alpha_{CU})t_{CU}$. In order to reduce the overall timing cost of the storage element, it is desirable to minimize $D_{DQ}(t_{CU})$, that is, maximize α_{CU}, for a given clock uncertainty.

Early Data Arrival   At this point it is worth mentioning that any problem with a long path delay can be fixed by reducing the clock frequency. Unlike the case of late data arrival, if the data arrives too early to be captured safely, the clocked system fails to operate correctly at any frequency. This is why assuring that the data arrives late enough to secure correct operation (meeting the *fast path requirement*) is one of the most critical issues in the design of any synchronous system. As expected, clock uncertainties make this task even harder.

When the clock uncertainties are present, the clock may arrive early at the source storage element of the pipeline stage and late at destination storage element (Fig. 5.5). Consequently, the data released from the source stage can arrive at the destination stage early enough to corrupt the previously captured data, creating a hold-time violation. The net effect of clock uncertainty is that the minimum delay of the fast paths in the logic has to be increased even further.

The impact of the clock uncertainty on early data arrival is illustrated on the example of the pipeline stage shown in Fig. 5.5. Clocks Clk_B and Clk_C are generated from the common clock Clk_A. The timing of the early-arriving data is associated with the same clock edge at both the source and destination storage element. Actual clock uncertainty that affects the path is the delay between the early arrival of Clk_B and late arrival of Clk_C. Consequently, any clock uncertainty of Clk_A affects both Clk_B and Clk_C in the same way, and thus has no influence on the fast path.

In order to avoid the hold-time violation, the sum of the minimum clock-to-output delay of the source storage element and logic delay must be greater than the hold time, under the most pessimistic assumption on clock uncertainty:

$$D_{CQm} + D_{Lm} \geqslant H + t_{CU} \qquad (5.7)$$

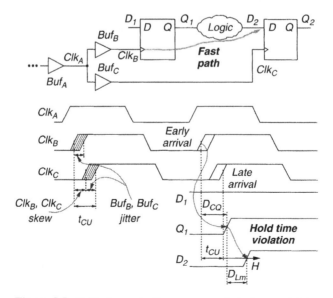

Figure 5.5. Critical race in the presence of clock uncertainty.

where D_{CQm} and D_{Lm} are minimum clock-to-output delay and minimal logic delay, respectively, H is the hold time of the destination storage element, and t_{CU} is the clock uncertainty between the clocks at the source and destination storage elements.

Components of the clock uncertainty that affect the fast paths are clock skew and clock distribution jitter generated within the clock domain that contains both of the path storage elements. We refer to this component of the clock uncertainty as *local clock skew* and *local clock distribution jitter*. Assuming the clock at both the source and destination storage elements are supplied by the same clock generator, clock generator jitter does not affect fast paths. Therefore, placing the source and destination storage elements on the path within the same clock domain is beneficial both in terms of minimizing skew impact on the slow-path requirement and skew and jitter impact on the fast-path requirement. In practice, it is hard to characterize local clock distribution jitter for all fast paths, so total jitter of the clock generator and clock distribution system, or clock distribution jitter only, can be used instead.

5.1.3. Clock-Uncertainty Absorbing Considerations

In order to achieve high clock-uncertainty absorption, the $D-Q$ delay characteristic of the storage element should be as constant as possible (*flat*) in the clock uncertainty window, as shown in Fig. 5.6. The nominal clock and data arrival times are 0 ps and -30 ps, respectively. If the clock triggering edge arrives -30 ps and 30 ps within the nominal clock arrival time, the output, Q, will not be affected. The output, Q, will still be generated 238 ps after the nominal data arrival (268 ps after nominal clock). In such a case, the output is not affected by the clock uncertainties.

The question is: How does one design a storage element with a flat data-to-output delay characteristic? This can be achieved by expansion of the time

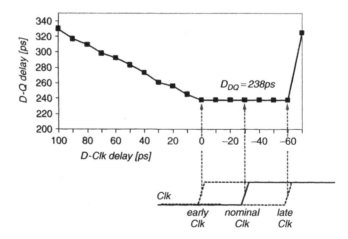

Figure 5.6. Idealized $D-Q$ delay characteristic as a function of clock arrival.

window during which the storage element is transparent (transparency window). Widening the transparency window is equivalent to increasing the separation between the two reference events in time: one that opens and other one that closes the CSE. In effect, the storage element behaves like a transparent latch for a short amount of time after the active clock edge. The wider the transparency window, the wider the flat region of the $D-Q$ characteristic, as described in Section 2.2.1. Widening the transparency window can be done by intentionally creating a wider capturing pulse of the flip-flops and pulsed latches, or overlapping the master and slave clocks of the MSLs.

A consequence of increasing the transparency window is that the failure region of the $D-Q$ characteristic is moved away from the nominal clock edge. This results in a decrease in setup time (larger negative values) and an increase in hold time of the storage element. While decreasing the setup time has no significant effect to the system timing as long as the $D-Q$ delay is constant, a long hold time makes the fast-path requirement harder to meet (Eq. (5.7)). Thus, the design for clock-uncertainty absorption is often traded for a longer hold time. In many cases, however, these two requirements are not contradictory, since a different type of storage element is used in the fast and slow paths.

5.2. TIME BORROWING

In a pipeline with level-sensitive clocking, the data input to the latch nominally arrives when the latch is transparent. A beneficial property of such a system is that a stage can use more time than nominal to produce its outputs, as long as this is compensated for by the subsequent (faster) stages. The technique of exploiting this property is called *time borrowing (cycle stealing, slack passing)* (Partovi et al. 1996; Harris and Horowitz 1997; Harris et al. 1996; Lin et al. 1992; Sakallah et al. 1992). A benefit of this technique is that the maximum clock frequency is obtained as an average of all stage delays, rather than the maximum delay of the largest stage delay, as with a pipeline with an edge-sensitive clocking. This level-sensitive clocking property avoids the increase in the cycle time caused by unbalanced logic delays between the pipeline stages. In this book, the type of time borrowing where the time borrowed is determined by the logic delays in the pipeline stage is called *dynamic time borrowing*.

The essential condition for logic in one pipeline stage to borrow time from another pipeline stage is that there are no "hard" boundaries between stages, that is, the storage elements are transparent at the time when data arrive. This transparency occurs in two clocking styles, level-sensitive and soft-edge clocking. Dynamic time borrowing is first discussed in Section 5.2.1 in the example of level-sensitive system using transparent latches. Then, in Section 5.3, we address the potential of using soft clock-edge flip-flops for dynamic time borrowing.

Another type of time borrowing is when the clock is intentionally delayed by inserting delay between the clock inputs of the clocked storage elements. The clock delays are scheduled so that the critical paths obtain more time to evaluate (the destination storage element captures data later), which takes time away from the faster paths. This technique is called *opportunistic skew scheduling*, and it is described in Section 5.2.2. Opportunistic skew scheduling statically assigns the maximum evaluation time to a stage by allowing for fixed additional time between the releasing and receiving clock. In this book we classify opportunistic skew scheduling as *static time borrowing*.

5.2.1. Dynamic Time Borrowing

A pipeline using two-phase level-sensitive latches is shown in Fig. 5.7a. Stages 1a and 2b are the logic blocks positioned between latches L_1-L_5. The latches are clocked by nonoverlapping clock phases Φ_1 and Φ_2. Timing diagrams for two stages of the pipeline are shown in Fig. 5.7b. Labels d_1 and d_2 represent the data flowing through the pipeline. Each logic stage alters the data according to pipeline functionality, and the labels only intend to show the signal propagation,

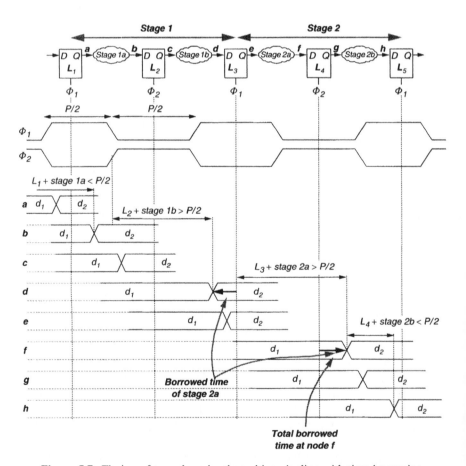

Figure 5.7. Timing of two-phase level-sensitive pipeline with time borrowing.

not the actual values. We assume that the latches are transparent on the "high" level of the controlling clock phase.

The borrowed time of a stage is the time difference between the actual and nominal stage delay. The *total* (or *accumulated*) borrowed time at any latch input is the time difference between the actual and nominal arrival of the latch input. We assume that the data nominally arrive at the input of a latch in the middle of the transparency period of the latch.

In the case of ideal logic partitioning, the delay of each stage should be half a clock cycle minus the delay of the latch. However, it is not always possible to partition the logic perfectly. In this example, *Stages 1a* and *2b* have a delay that is smaller than ideal, *Stages 1b* and *2a* have a delay that is larger than ideal, and the input data to the pipeline arrive prior to the leading edge of the clock, Φ_1. *Stage 1a* receives the data after the leading edge of the clock, Φ_1, and produces an output before the leading edge of the clock, Φ_2. Consequently, signal c does not change until the arrival of the leading edge of Φ_2. Since *Stage 1b* introduces a delay larger than half the clock period, signal d arrives during the transparency period of L_3, resulting in signal e (after propagating through L_3). Even though *Stage 1b* takes more time than nominally assigned, signal d still arrives prior to its nominal arrival. Thus, *Stage 1b* borrows time from *Stage 1a*, but the total borrowed time at signal d is negative. *Stage 2a* borrows time from both *Stages 1b* and *2b*, so that signal f arrives after the middle of the transparency period of L_4. The borrowed time is "returned" the (total borrowed time is negative at signal h) in *Stage 2b* and signal h arrives before its nominal arrival time. If the delay of *Stage 2a* were larger than shown in Fig. 5.7, the setup time of latch L_4 may have been violated and the pipeline would not operate correctly.

The preceding example illustrates several key issues of time borrowing:

- The maximum throughput is not determined by the worst-case delay of the slowest logic block (*Stage 1b* and *Stage 2a*), but rather by the average delay of all of the pipeline stages.
- Only the stages that receive or deliver data through a transparent latch participate in time borrowing. The time between the arrival of signal b and the moment latch L_2 becomes transparent is not used from the perspective of time borrowing (Fig. 5.7).
- The borrowing cannot continue indefinitely; in any event, the data must arrive early enough to be captured by the subsequent latch. At any latch input in the pipeline, accumulated borrowed time must not exceed the value at which the setup time of the latch is violated. Thus, for the nominal data arrival in the middle of the transparency period, the maximum accumulated borrowed time is half the transparency period reduced by the setup time of the destination latch of the stage.

Timing Analysis with Time Borrowing

Late Data Arrival   For any latch in the system that exploits time borrowing, the data must arrive in time to be properly captured. The data arrival at the latch is a function of the delays of all previous pipeline stages. Timing of all the stages that share a logic path with a pipeline stage must be calculated in advance in order to obtain worst-case data arrivals for that stage. The cases where loops exist may require complex iterative procedures, since no data arrivals are known initially. All this makes slow-path analysis more complex than that of a pipeline synchronized by flip-flops.

In order to see how time borrowing affects minimum clock cycle time, consider a system of N pipeline stages, each divided by transparent latches into two logic blocks, as shown in Fig. 5.7a. The latches are controlled by clock phases Φ_1 and Φ_2. All logic blocks are used in time borrowing, that is, worst-case data arrival occurs only when the latch is transparent. Thus, the arrival time of the input at the subsequent latch, $t_{D,i+1}$ is equal to the sum of the arrival times of the input to the preceding latch, $t_{D,i}$, the $D-Q$ delay of the latch, $D_{DQ,i}$, and the logic delay, $D_{LM,i}$, of the current stage:

$$t_{D,i+1} = t_{D,i} + D_{DQ,i} + D_{LM,i} \tag{5.8}$$

The arrival of input at the $(2N + 1)$-th latch (input at the $(N + 1)$-th stage) is

$$t_{D,2N+1} = t_{D,1} + \sum_{i=1}^{2N} (D_{DQ,i} + D_{LM,i}) \tag{5.9}$$

We assume that the after N stages, the pipeline produces data at the same point in the clock-phase, Φ_1, transparency period at which the input data was acquired in the first clock cycle. Therefore, $t_{D,2N+1} - t_{D,1}$ is equal to N clock periods P:

$$t_{D,2N+1} - t_{D,1} = NP \tag{5.10}$$

Combining Eq. (5.9) and Eq. (5.10), we obtain the requirement for the minimum clock period under late data arrival:

$$P = \frac{1}{N} \sum_{i=1}^{2N} (D_{DQ,i} + D_{Logic,i}) \tag{5.11}$$

Equation (5.11) shows that the minimum clock cycle time of the pipeline is *not* determined by the delay of the slowest stage in the pipeline. It is rather the *average* delay of the logic and latches through all stages. Thus, the speed-up can be achieved by giving slow stages more time to evaluate at the expense of faster stages. Note that Eq. (5.11) is valid only if the data arrive at the latch during the time it is transparent. This important constraint has to be true for all latches on the path, and is summarized in the following equation:

$$\frac{i-1}{2} P - D_{DQ,i} + D_{CQ,i} < t_{D,i} < \frac{i}{2} P - U_i, \qquad \forall i \in [1..2N] \tag{5.12}$$

where it is assumed that the first leading edge of Φ_1 occurs at time zero, and U_i and $D_{CQ,i}$ represent the setup time and clock-to-output delay of latch i, respectively.

Early Data Arrival   In a time-borrowing system, fast paths can cause the pipeline to operate incorrectly. When the destination latch is still transparent during the time the source latch becomes transparent, a short path in the logic can cause the latching of the data from the same clock cycle. The clock uncertainties make the system more vulnerable to hold time failures. In order to ensure the correct operation of the pipeline, it should be provided that the minimum stage delay exceeds some specified value.

In order to illustrate the effect of time borrowing on the fast paths, we refer to Fig. 5.8, which is an excerpt of Fig. 5.7. Signal d changes while L_3 is still opaque. When Φ_1 rises, L_3 becomes transparent and signal e changes after a latch delay. The change in signal e propagates to signal f after the delay of *Stage 2a*. In the case where the sum of the L_3 delay and the delay of *Stage 2a* are larger than the sum of the overlap of Φ_1 going high to Φ_2 going low and the hold time of L_4, the data will race through both L_3 and L_4 in one phase of the clock, causing a functional failure. Thus, the time-borrowing technique does not help alleviate fast-path hazards, so these hazards should be treated as discussed in Chapter 4, that is, assuming no time is borrowed between the stages. The only remedy to the fast-path problem is either to make Φ_1 and Φ_2 strictly nonoverlapping, or to pad every fast path in the pipeline with extra logic to guarantee some minimum required logic delay.

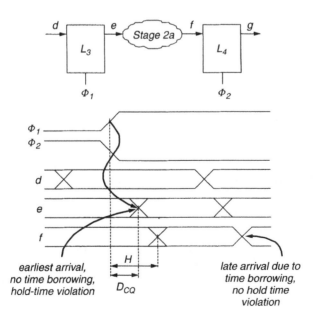

Figure 5.8. Fast-path hazard.

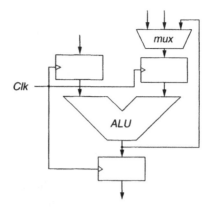

Figure 5.9. Forwarding path in a pipeline.

Loop Requirement In time-borrowing systems, the timing of signals in the loops (feedbacks), which are commonly employed in the pipelines, should be treated separately from other paths. An example of such a loop is the forwarding path that feeds the data from the output of the execution stage back to its input in order to prevent the pipeline hazards (Fig. 5.9). If the overall propagation delay through the loop consisting of N stages exceeds NP (where P is the clock period), the arrival time may occur later with each cycle, finally resulting in a setup time violation. More generally, any signal loop that borrows time from itself will eventually cause a timing violation.

5.2.2. Static Time Borrowing

The static time-borrowing technique, often referred to as *opportunistic skew scheduling* or *optimal skew scheduling* (Fishburn 1990; Friedman 1995), exploits intentional delay insertion between clock inputs of different storage elements. In this way, evaluation time per stage can be better distributed by giving additional time to slow stages at the expense of the fast ones. This technique is applicable to the systems in which there are stages that use less time for computation than allocated by the clock cycle.

A typical opportunistic skew scheduling scheme is shown in Fig. 5.10. Each clocked storage element, CSE_i, in the system receives the reference clock, delayed by time Δ_i. The clocks are distributed in such a way that the storage elements preceding the longest paths in combinational logic receive the early clock, and the storage elements following the longest paths receive the delayed clock. For example, Clk_2 in Fig. 5.10 is delayed for Δ with respect to Clk_1, so that slower *Stage 1* is allocated more time at the expense of faster *Stage 2*. Consequently, the system can be clocked at a higher rate than what would otherwise be dictated by the delay of the slower *Stage 1*.

A benefit of static time borrowing is that it can operate with conventional flip-flops. In addition, it places fewer constraints on the circuit design, allowing

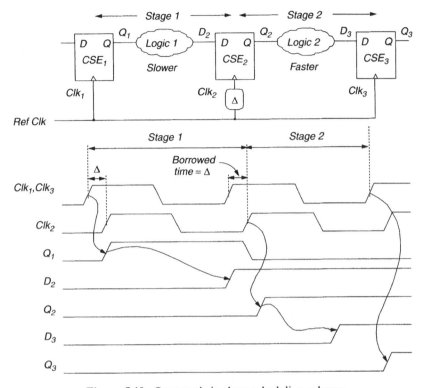

Figure 5.10. Opportunistic skew scheduling scheme.

longer critical paths where necessary. This very appealing concept of static time borrowing has a few disadvantages. It increases the complexity of the clock distribution system. In particular it is hard to control the inserted delays over process, supply, and temperature variations. Also the analysis of clock skew is complicated in this asymmetric clock distribution network. While all these difficulties make this technique impractical on a large-scale level, it is nonetheless very useful in localized critical paths where every improvement directly increases the system clock rate.

In conclusion, it is important to notice the difference between dynamic and static time borrowing. In dynamic time borrowing, the transparency of the storage element itself is exploited and the time is borrowed based on the actual differences in the stage delays. Consequently, the amount of borrowed time depends on the delay of the logic blocks in the stages. Also, the clock can be distributed uniformly.

5.3. TIME BORROWING AND CLOCK UNCERTAINTY

Both the clock-uncertainty-absorption and dynamic-time-borrowing techniques use the storage-element property to reduce the effect of indeterminate data-to-clock delay to data-to-output delay. This single property can be interpreted in two apparently different ways. While for clock-uncertainty absorption, indeterminate data-to-clock delay is caused by the uncertainty of clock arrival, for time borrowing it is caused by uncertain data arrival. In both cases, the transparency of the storage elements is used to suppress the input uncertainty (either that of the clock or the data). Thus, clock-uncertainty absorption and time borrowing are essentially equivalent properties. If a clocking strategy allows dynamic time borrowing between the stages, it will also be capable of absorbing the clock uncertainty, and vice versa. In Section 5.3.1, we address this analogy by describing the uncertainty-absorbing capability of a level-sensitive latch-based system, whose time-borrowing property is discussed in Section 5.2. Subsequently, in Section 5.3.2 we show that the soft-clock-edge property of the flip-flops, which is responsible for clock uncertainty absorption (Section 5.1), can be used for time borrowing between the stages.

5.3.1. Level-Sensitive Clocking

In Section 5.2 we saw that time borrowing exploits the data arrivals in the latch transparency period to allow more time for logic evaluation. Equivalently, if the data arrive during the latch transparency period, the actual moment of clock arrival does not affect the timing of the signals in the pipeline. This means that the slow-path timing relation in the pipeline stage and minimum clock period are immune to the clock uncertainties. If we are able to keep all data arrivals in the middle of the transparency period of the capturing latch, all pipeline stages, and therefore the system as a whole, would be immune to the clock uncertainty (up to about half of the latch transparency period).

The essential condition for allowing the pipeline to absorb the clock uncertainty is that the data arrive at the latch input while the latch is transparent. For example, if the Φ_2 clock controlling latch L_4 in Fig. 5.7 arrives a little earlier or later than shown in Fig. 5.7b, the rest of the timing diagrams will not change. This is because the data arrive while latch L_4 is transparent. However, this is not true for latch L_2, since it must wait for the clock in order to release data c to the subsequent stage. Consequently, any fluctuation in the arrival of clock Φ_2 is passed onto signal c.

Timing Analysis

Late Data Arrival   To determine the level of clock uncertainty tolerable to the latch-based time-borrowing system, we refer to Fig. 5.11. The figure illustrates the timing relationship between clock (Φ_1) and the data (D) input to the latch that is part of the time-borrowing pipeline (Fig. 5.7). It is assumed that the latch is transparent during the high level of the clock and that the nominal duration of the high level of the clock is W. The latch input arrives later than the clock for the amount of time t_D. This is equivalent to time borrowing of $t_B = t_D - W/2$ at signal D, assuming the nominal arrival of D is in the middle of the transparency period of the latch. We use T_L and T_T to denote the maximum uncertainties of the

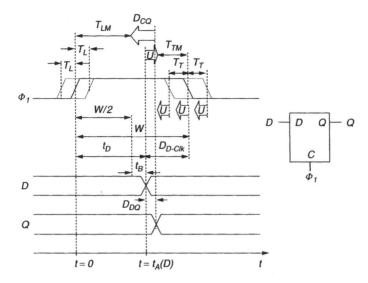

Figure 5.11. Clock uncertainty immunity in the pipeline with level-sensitive clocking.

leading and trailing edges of the clock, respectively. The clock-to-output delay and D–Q delay of the latch are D_{CQ} and D_{DQ}, respectively. Signal arrival with respect to the nominal arrival of the clock is designated t_A.

As long as the leading edge of the clock arrives early enough so that the latch is still transparent at the time the data arrive, the latch output and the rest of the signals in the pipeline do not change. The latest arrival of the leading edge of the clock that does not affect the pipeline timing is determined by Eq. (5.13):

$$t_A(\Phi_1) + D_{CQ} > t_A(D) + D_{DQ} \tag{5.13}$$

Since the leading edge of the clock nominally arrives at time $t = 0$, its latest arrival time is T_L. Consequently, the bound for the uncertainty of the clock's leading edge is

$$T_L \leqslant t_B + D_{DQ} - D_{CQ} + W/2 \tag{5.14}$$

Any value of the late clock arrival time smaller than the bound in Eq. (5.14) does not affect the timing of the output of the latch.

An increase in early arrival of the trailing edge of the clock does not have an effect on the pipeline timing, as long as the data arrive at setup time, U, before the trailing edge of the clock:

$$D_{D\text{-}Clk} - T_T \geqslant U \tag{5.15}$$

where $D_{D\text{-}Clk}$ is the time between data arrival and the trailing edge of the clock. From Fig. 5.11, this $D_{D\text{-}Clk}$ is equal to $W/2 - t_B$. Thus,

$$T_T \leqslant W/2 - t_B - U \tag{5.16}$$

As long as the clock uncertainty satisfies the inequalities Eqs. (5.14) and (5.16), the maximum throughput is not affected, because Eq. (5.11) holds and the time borrowing can be exploited. The effect of the clock uncertainty on time borrowing can be observed by rewriting Eq. (5.16):

$$t_B \leqslant W/2 - U - T_T \tag{5.17}$$

Equation (5.17) shows that the early arrival of the trailing edge of the clock caused by the clock uncertainty reduces the maximum allowable amount of time borrowing. Similarly, Eq. (5.14) shows that the late arrival of the leading edge of the clock reduces the time that the stage can accumulate for later borrowing (if the nominal data arrival time is less than $W/2$). Thus, we are trading off time borrowing for tolerance to the clock-edge uncertainty.

Ideally, time borrowing should be extended to all pipeline stages so that the input to each latch arrives when the latch is transparent. However, this cannot be accomplished for all stages. For example, the time when the latch in front of the first stage of the pipeline receives its input is specified at the system level and it cannot be chosen by the circuit designer. We can estimate the overall effect of the clock uncertainties on the pipeline if we observe a multicycle critical path shown in Fig. 5.12. The path starts from latch L_1 and ends at latch L_7. Latches $L_2 - L_6$ receive their inputs around the middle of their transparency period. Therefore, the setup time of latches $L_2 - L_6$ is satisfied and the clock uncertainties of clock phases Φ_1 and Φ_2 have no effect on timing. Since latch L_1 waits for Φ_1 to release the data, the clock uncertainty of Φ_1 reflects on Q_1 and propagates through the critical path to the end of *Stage 3*. The setup time margin of latch L_7 is reduced

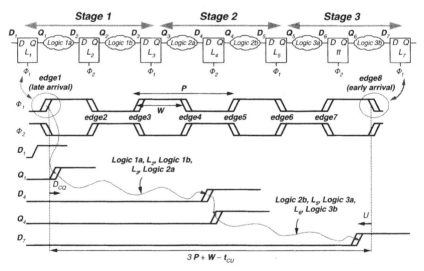

Figure 5.12. Impact of clock uncertainties on the critical path in the pipeline with time borrowing.

if the trailing edge, *edge8*, of clock Φ_1 arrives early. In order to meet the setup time requirement of latch L_7, the following must hold:

$$t_{CU} + D_{CQl} + D(D_1 \rightarrow D_7) + U \leqslant 3P + W \tag{5.18}$$

Equivalently,

$$P \geqslant \frac{D_{CQl} + D(D_1 \rightarrow D_7) + U - W}{3} + \frac{t_{CU}}{3} \tag{5.19}$$

In Eqs. (5.18) and (5.19), D_{CQl} is the clock-to-output delay of L_1 and $D(Q_1 \rightarrow D_7)$ is the delay of the path from Q_1 to D_7. The clock uncertainty, t_{CU}, is the uncertainty of the trailing edge of the clock at L_7 with respect to the leading edge of the clock at L_1. Equation (5.19) shows that the impact of the clock uncertainty on the minimum clock period is reduced by being divided among the number of stages that the critical path goes through. The source latch and destination latch of the critical path (latches L_1 and L_7 in Fig. 5.12) are normally placed in the same clock domain, thus reducing the clock skew between them. Note that, since several clock edges occur during the evaluation of the critical path, the clock jitter between *edge1* and *edge8* is larger than the cycle-to-cycle clock jitter used in single-stage analysis. This observation is true for all systems that absorb the clock uncertainties, since in all such systems the timing in a pipeline stage depends on the data arrivals from previous clock cycles.

Early Data Arrival The direct effect of the clock uncertainties on the fast-path requirement in the multiphase level-sensitive pipeline is that the overlap between the phases increases. If, for example, in Fig. 5.12 the leading edge of clock phase Φ_1 arrives early, and/or the trailing edge of clock phase Φ_2 arrives late due to the uncertainty, the overlap between the phases is the sum of the two clock uncertainties. As this overlap increases, the fast path, discussed in Section 5.2, can cause erroneous operation. The earliest arrival of the clock's leading edge and the latest arrival of the clock's trailing edge that the system can tolerate are set by the hold time requirement:

$$D_{CQm} + D_{Lm} \geqslant (V + T_{L,\Phi_1} + T_{T,\Phi_2}) + H \tag{5.20}$$

$$T_{L,\Phi_1} + T_{T,\Phi_2} \leqslant D_{CQm} + D_{Lm} - V - H \tag{5.21}$$

In Eqs. (5.20) and (5.21), T_{L,Φ_1} and T_{T,Φ_2} designate the early arrival of leading edge of clock Φ_1 and the late arrival of the trailing edge of clock Φ_2, respectively, and V is the nominal overlap between Φ_1 and Φ_2. Equation (5.21) provides a conservative rule for making the fast paths robust to the clock uncertainties.

In summary, the influence of the clock uncertainties on the timing in a time-borrowing system is in:

- Decreasing of the margins for time borrowing. Both the minimum allowed path delay and the maximum allowed time borrowing are reduced by the clock uncertainty.

- The pipeline absorbs the uncertainties for the data that arrive during the latch transparency period.

- The effect of the uncertainties is reduced to an average uncertainty over all stages in the path.

5.3.2. Soft-Edge-Sensitive Clocking

The clock uncertainty absorption, α_{CU}, defined in Section 5.1.1, shows how the propagation delay of a flip-flop is changed if its clock timing is uncertain. Applying this clock uncertainty to a flip-flop is equivalent to keeping its clock arrival fixed and allowing data arrival to change. Thus, more generally, the parameter α_{CU} quantifies the degradation of the $D-Q$ delay for uncertain data-to-clock delay. As such, it can be used to describe the timing of the flip-flop if it is used in time borrowing in exactly the same way it is used for clock-uncertainty

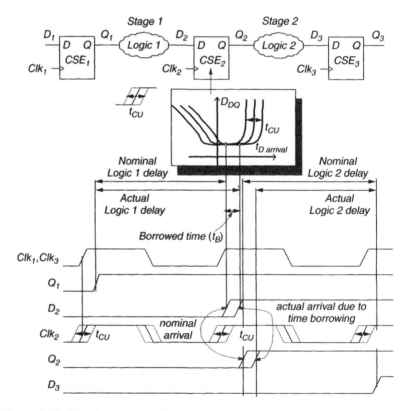

Figure 5.13. Time borrowing with uncertainty-absorbing clocked storage elements.

absorption. In this perspective, high α_{CU} (soft clock edge) designates a storage element whose output follows both the early and late input arrivals, allowing slower stages to borrow time from the subsequent faster stages.

The time-borrowing capability and the clock-uncertainty absorption are not mutually exclusive. In fact, they can be traded off for each other. Figure 5.13 illustrates a case where a wide transparency window, denoted as a flat $D-Q$ characteristic, is used to both absorb the clock uncertainties, t_{CU}, and borrow time, t_B, from the surrounding stages. Combinational logic of *Stage 1* takes more time than nominally assigned, and it borrows a portion of the cycle time from *Stage 2*. In general, the storage element may not be completely transparent (i.e., the $D-Q$ characteristics are not completely flat). According to the definition of clock-uncertainty absorption, the combination of clock uncertainty, t_{CU}, and time borrowing, t_B, causes an increase in the $D-Q$ delay of the flip-flop, ΔD_{DQ}:

$$\Delta D_{DQ} = (t_B + t_{CU})[1 - \alpha_{CU}] \qquad (5.22)$$

where α_{CU} is a function of $t_B + t_{CU}$. The delay increase, ΔD_{DQ}, is the same either when the clock uncertainty is $t_B + t_{CU}$ with no time borrowing, or when the borrowed time between stages is $t_B + t_{CU}$ and there is no clock uncertainty.

It should be noted that the practical values of the total borrowed time are similar to the width of the transparency window, and in any event are shorter than the hold time. Better absorption and time-borrowing capability can be obtained by widening the transparency window (see Section 5.1.3). However, if the transparency window is widened, the hold time increases and the short-path requirement becomes harder to meet. Therefore, use of a wide transparency window is a trade-off between time borrowing and uncertainty absorption on the one side and the hold time on the other side. In cases where sufficient minimum delay in the logic path can be assured, making this window wider can be beneficial.

CHAPTER 6

LOW-ENERGY SYSTEM ISSUES

A large portion of the energy consumption in modern microprocessor designs is in the clock subsystem, including clock generation, distribution, and the final clocked storage-element load. Due to increasing frequency, low skew requirements, and deep pipelining, this clocking energy has been increasing with each processor generation, requiring a more energy-conscious design of the clock subsystem. In this chapter we describe some widely used methods for energy reduction that include supply-voltage scaling, minimizing switched capacitance, minimizing switching activity, and the use of low-swing-circuit techniques. These conventional principles are then applied to the design of alternate topologies of clocked storage elements as well as a general clock distribution network.

A common design approach for minimizing energy consumption in VLSI systems is to concentrate on reducing the switching component of energy, given by

$$E_{\text{switching}} = \sum_{i=1}^{N} \alpha_{0-1}(i) \cdot C_i \cdot V_{\text{swing}}(i) \cdot V_{DD} \qquad (6.1)$$

where N is the number of nodes in the system, C_i is the capacitance at node i, $\alpha_{0-1}(i)$ is the probability that the energy-consuming transition occurs at node i, V_{swing} is the voltage swing of node i, and V_{DD} is the global supply voltage. Based on this simple formula, the guidelines for reducing energy consumption are simply to minimize each of the terms in the product expression. The most efficient way to minimize energy, as should be obvious, is aggressive voltage scaling, because the energy has, to the first order, a quadratic dependency on the supply voltage.

Supply-Voltage Scaling Energy consumption is a quadratic function of the supply voltage, so operating at reduced supply voltages leads to significant savings in energy consumption. In digital systems that deal with supply voltage scaling (Burd et al. 2000), it is desirable that this scaling be used to preserve important timing relationships. In particular, it is important that delays of both the clocked storage elements and combinational logic scale in the same way to maintain the timing constraints imposed by the fast and slow paths without any changes in the design. While further slowing down the slow paths only affects the maximum clock rate, speeding up the fast paths in order to avoid critical races is not acceptable. Figure 6.1 illustrates the delay and internal race immunity of representative MSLs, flip-flops, and locally gated latches and flip-flops (Markovic et al. 2001).

In general, flip-flops (HLFF, SDFF, M-SAFF) are desirable circuits for critical paths at reduced supply voltage because the delay of these elements becomes shorter at lower supplies, relative to the delay of a static CMOS FO4 inverter, as illustrated in Fig. 6.1a. This is because of the favorable scaling of the stack transistors in this particular technology, as illustrated in the example of the two-input NAND gate. With scaling down the supply voltage, the equivalent threshold voltage of the NAND gate decreases, due to the reduced impact of the body effect. This outweighs the threshold increase due to DIBL effect at reduced supply voltage, resulting in overall relative speed-up. However, this behavior is dependent on the underlying technology, and should not be taken as a general rule. Circuits with transistor stacks (HLFF, SDFF) showed behavior similar to that of the NAND gate, while M-SAFF had the largest speed-up due to its cross-coupled differential structure with positive feedback circuits. Unlike the delay, the internal race immunity of these latch and flip-flop topologies does not change with supply voltage relative to an FO4 inverter (Fig. 6.1b), indicating that the same fast-path constraints apply across a range of supply voltages. This is because Clk-Q delay and hold-time scale in a fashion similar to that of a CMOS inverter, due to the nature of the circuits that define Clk-Q delay and hold time.

Minimizing Effective Switched Capacitance In order to obtain maximum energy savings, the goal is to minimize all the effective switched capacitance internal to the clocked storage element, for a given external load capacitance. The total effective capacitance at some node inside the circuit is a product of the physical capacitance of that node and probability of the energy-consuming transition. Physical capacitance includes clocked transistor capacitance and capacitance of nonclocked internal nodes. Reduction of the total physical clocked capacitance is more important because it is switched every clock cycle, as opposed to the capacitance of the nonclocked nodes, which is switched only when the output changes. In dynamic circuits, however, only a portion of the nonclocked capacitance — the total capacitance excluding the capacitance of the precharge/evaluate nodes — is switched when the output changes.

Circuit Sizing One advanced issue in minimizing energy consumption in clocked storage elements pertains to the circuit sizing that provides an optimal

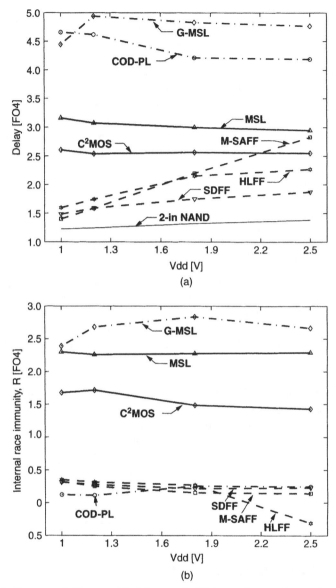

Figure 6.1. Impact of V_{dd} on (a) delay, and (b) internal race immunity (0.25 μm, light load). (Markovic et al. 2001), Copyright © 2001 IEEE.

energy-performance trade-off for a given output load capacitance. Ideally, we would like to have the lowest possible energy and the highest level of performance, but the two requirements conflict. Intuitively, clocked nodes should be made minimum size in order to compensate for the increased switching activity. The total circuit area ultimately depends on the size of the load that it needs to drive, implying that larger loads may need a design of larger proportions to maintain acceptable driving strength. However, it is not desirable to size transistors in such a way that they are overly robust. This means that a circuit with transistors of fixed size cannot optimally drive various output loads, and that the extra area spent in designing them to support the largest load capacitance is really wasted, since the performance upgrade here serves only to alleviate an issue that may manifest itself infrequently. It is more important from an energy standpoint that circuits are sized to satisfy the constraints in the most common cases, which often happens to be approximately a fourfold increase over standard inverter load. Typically, one "standard load" corresponds to the input capacitance of a "1×" buffer from the standard cell library. In more advanced VLSI designs today, the output loads are even lower 80% of the time.

Minimizing the physical capacitance by downsizing the transistors is often limited by the requirements for circuit noise immunity. For this reason, for example, standard cell libraries typically do not contain minimum-sized transistors at the inputs of logic gates. It is important to optimize the size of a clocked storage element for minimal energy that just meets the performance goal. There is no type of storage element that is optimal for all paths. Performance critical paths require fastest operation, which results in sizing for peak performance, and thus large energy consumption. Storage elements in noncritical paths allow much less aggressive sizing due to the available timing slack. Circuit-size optimization thus depends on the topology of a clocked storage element and is discussed in more detail in Chapter 7.

Circuit Topology   Other efficient ways to minimize the overall energy are lowering clock signal swing and reducing clock frequency. Selection of a circuit style that has inherently low switching activity in the internal nodes or a small number of clocked nodes could also be a good way of reducing energy. For example, in most cases, static circuits have smaller energy consumption than dynamic circuits, because the dynamic circuits need precharge/discharge operation of the dynamic nodes in each clock cycle. In addition, effective switching activity can be reduced by clock gating or dual edge-triggering on every clock transition that halves the frequency of the global clock.

6.1. LOW-SWING CIRCUIT TECHNIQUES

Clocked storage elements sometimes operate with different input- and output-signal logic levels. For example, in static random-access memories (SRAM), a low-swing wordline signal is amplified by sense amplifiers to produce a full swing at the bus output. If the data and clock loads are similar, then it is much more beneficial from the energy standpoint to have a low-swing clock. This is because of the high proportion of the clock energy to data energy due to the high switching activity of the clock. Another low-swing approach is therefore the reduced-swing clock operation, which targets savings in the clocking energy. The low-swing clock can be generated with the help of reduced-swing clock drivers or by powering up the clock buffers with a separate supply voltage (Kawaguchi and Sakurai 1998). The aim of the reduced-swing clocking technique is to save energy in the clock network. The reduced-swing clock operation slows the circuit down and consequently increases the cycle time. This technique is therefore effective only in VLSI systems where an increase in the clock cycle is allowed. Since supply voltage has a stronger effect on the delay than circuit sizing, upsizing the circuit that operates with reduced-swing signals usually cannot average out the performance loss experienced from the low-swing operation. Therefore, there is always some delay penalty associated with reduced-swing signals. Low-swing clocking can be implemented either with conventional CSEs and specially designed clock drivers, or with specially designed CSEs that are capable of receiving the reduced-swing clock.

6.1.1. Conventional CSEs with Reduced-Swing Clock Drivers

When used with conventional CSEs, low-swing clocking requires a special design for the clock drivers to support the reduced-swing clock operation. As an example, consider the clock driver proposed by Kojima et al. (1995). It provides separate clock signals for p-MOS and n-MOS transistors, as shown in Fig. 6.2. Capacitances C_{p1}, C_{p2} represent p-MOS loads on the driver, and C_{n1} and C_{n2} represent n-MOS loads on the driver. Capacitors C_A and C_B are externally connected or fabricated on-chip, to optimally set voltage at node H-V_{DD} to $V_{DD}/2$:

$$V(H\text{-}V_{DD}) = \frac{C_{p1} + C_A}{C_{p1} + C_{n2} + C_A + C_B} V_{DD} \quad (Clk \text{ is low}) \quad (6.2)$$

$$V(H\text{-}V_{DD}) = \frac{C_{p2} + C_A}{C_{p2} + C_{n1} + C_A + C_B} V_{DD} \quad (Clk \text{ is high})$$

When C_{p1}, C_{p2}, C_{n1} and C_{n2} are made equal, then the node H-V_{DD} is at $V_{DD}/2$ and the external capacitors C_A and C_B are not needed. Otherwise, C_A and C_B can be made large in a way that makes variations in C_{p1}, C_{p2}, C_{n1}, and C_{n2} insignificant and sets H-V_{DD} close to $V_{DD}/2$. Each of the clock buffers in Fig. 6.2 would provide a half-swing clock signal for p-MOS or n-MOS transistors. This

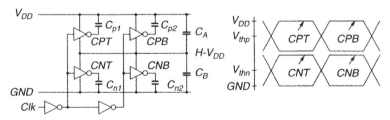

Figure 6.2. Clock driver for half-swing clocking. (Kojima et al. 1995), Copyright © 1995 IEEE.

way, both phases of the clock are provided and conventional topologies of clocked storage elements can be used.

Eliminating any cost in the clock driver associated with charging C_A, C_B and generating both phases of the clock, half-swing clocking ideally provides 75% reduction in the clocking energy. In reality, additional resizing of n-MOS clocked transistors is performed to balance $C_{p1}-C_{n2}$. As a result, smaller energy savings are achievable. For example, the sixteen-stage shift register reported in Kojima et al. (1995) saves 67% of the clocking energy, 8% less than the theoretical result. This technique is limited only to those cases where a half-swing operation is required. Alternate techniques dealing with redesign of the clocked storage elements allow an arbitrary value of the reduced clock swing to be used and more flexibility in optimizing the overall system energy.

6.1.2. CSE Redesign

Kawaguchi and Sakurai (1998) took a different approach to low-swing clocking. They supplied a globally reduced-swing clock to all clocked transistors, with extra body bias applied to all p-MOS clocked transistors. This is because they do not fully turn off when the *Clk* is high. Their study was based on the example of a sense-amplifier-based flip-flop (SAFF), modified for the reduced-swing clock operation, as shown in Fig. 6.3. Operational behavior of the reduced clock swing flip-flop (RCSFF) is very similar to the behavior of the SAFF. The RCSFF example is particularly interesting because the new circuit did not require any topological change, but rather the use of an extra bias voltage to bias wells of the precharge p-MOS transistors as a way of suppressing their leakage current when the clock is high.

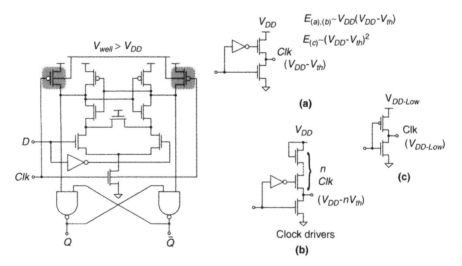

Figure 6.3. Reduced clock-swing flip-flop. (Kawaguchi and Sakurai 1998), Copyright © 1998 IEEE.

The reduced-swing clock can be either globally distributed or it can be generated locally. Kawaguchi and Sakurai proposed several low-swing clock drivers that reduce the output signal swing by stacking up n n-MOS transistors to generate an output swing of $V_{DD} - nV_{th}$, as shown in Fig. 6.3. However, a more effective method is to design a clock distribution network comprising standard buffers and globally reduced clock supply voltage. This is because particular schemes with stacked transistors result in increased pull-up resistance, and hence require some area overhead to maintain sharp clock edges and provide energy saving in the clock distribution network. Schemes (a) and (b) in Fig. 6.3 are less energy efficient than scheme (c), because the energy is effectively pulled out of V_{DD} and not V_{DD}-V_{th}, so the energy benefit is not quadratic, as with the scheme in (c).

Compared to the conventional MSL in Kawaguchi and Sakurai (1998), RCSFF provided 63% savings in the clocking energy for the same *Clk-Q* delay. We must note here that comparison with the conventional M–S topology is not quite appropriate because the M–S configuration is inherently slower than the SAFF. A fairer comparison would be with a SAFF that issues full-swing clocks, but whose *Clk* transistors are downsized such that the delay is equal to that of the RCSFF. Here, the RCSFF just serves to illustrate one of the few reduced-swing clock design options.

6.1.3. N-Only CSEs with Low-Supply-Operated Clock Drivers

The low-swing clock techniques discussed thus far are suboptimal. The technique presented in Section 6.1.1 that provides two different low-swing clocks for conventional flip-flops (Kojima et al. 1995) inherently increases the physical capacitance of the clock network. An approach with a separate well bias (Kawaguchi and Sakurai 1998) increases the layout complexity. A more effective technique is to use conventional clock drivers with a globally reduced supply voltage and CSEs containing only n-MOS-clocked transistors that are capable of receiving the low-swing clock. One such latch circuit is shown in Fig. 6.4.

This straightforward implementation is obtained by simply removing p-MOS clocked transistors from the conventional MSL. A design for robustness and speed involves adding extra n-MOS transistors $N_1 - N_4$ to help the pull-up transition on the latch state nodes S_M and S_S.

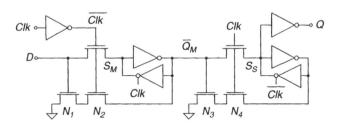

Figure 6.4. N-Only clocked M–S latch.

6.2. CLOCK GATING

Clock gating is an efficient way of reducing the overall energy consumption in digital systems where energy in the clocking subsystem is a significant part of the overall system energy, or when data input of the CSEs have little switching activity. The mechanism behind clock gating is to allow clocking of a CSE only if new arriving data are different from the current output of the CSE, which effectively eliminates switching of the clocked transistors when output does not transition. This way, unnecessary activity of the internal nodes is eliminated.

The clock gating can be *global* when the gating logic is shared between several CSEs, or *local* when the gating logic is embedded in each CSE. In both cases, the design of the clock-gating control logic needs to be carried out carefully so that the savings in the clocking energy are greater than the overhead incurred by the clock-gating logic, for the given input data statistics. Generally, extra caution has to be taken in the design of systems with gated clocks, because timing analysis becomes more complicated when the clock is gated (Baeg and Rogers 1999).

6.2.1. Global Clock Gating

Sometimes designers need to control which data are loaded into registers. To achieve this, an extra signal is needed to control the loading of new data into the registers. This kind of functionality can be essentially performed in two different ways: by employing a free-running clock and multiplexing (gating) the data, or by gating the clock signal. The standard way of recirculating data is shown in Fig. 6.5a. The circuit has a free-running clock and a "wrap-up" multiplexer that selects either the value stored at output *Q* or new input *In*. The selection is regulated by control signal *Load*, which represents the gating condition. The clock signal, *Clk*, triggers the register *REG* in each clock cycle. This is the more common approach, used in the LSSD methodology, and it prohibits insertion on the clock.

The principle of clock gating is illustrated in Fig. 6.5b. The circuit sends clock signal *Clk* to the register only when signal *EN* is active high. In this circuit, signal *EN* must not transition when signal *Clk* is high in order to assure capturing the input *In* at the edge of the clock signal *Clk* rather than at the edge

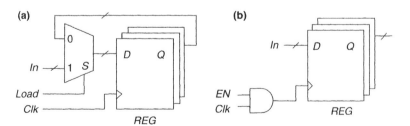

Figure 6.5. (a) Nongated clock circuit, (b) gated clock circuit. (Kitahara et al. 1998), Copyright © 1998 IEEE.

of the enable signal *EN*. In addition, there is always some extra insertion delay by the clock-gating logic. For these reasons, timing analysis in designs that employ global clock gating is more complicated than the timing analysis of conventional designs. This is a more energy-efficient method than the LSSD.

6.2.2. Local Clock Gating

Unlike globally gated designs, locally gated designs can be simply analyzed just like the conventional designs in which the clock-gating logic is lumped into a CSE. The main feature of this circuit family is a mechanism for predictive turn-off of the internal clock when the input and output data are equal. The local clock-gating technique can be applied to any CSE topology. The local clock gating in most cases incurs an extra delay penalty that effectively limits its applicability only to those CSEs that are outside performance-critical paths, such as in data-transition look-ahead latch (DTLA-L). However, there are cases when the control logic is outside the critical path of the CSE, in which case there is almost no penalty in the CSE delay. One example of such a design is the conditional capture flip-flop.

Example: Data-Transition Look-Ahead Latch   The DTLA-L proposed by Nogawa and Ohtomo (1998) is shown in Fig. 6.6. In the original paper, this circuit is called "flip-flop." The circuit is derived from a conventional-based MSL. It consists of the MSL, pulse generator, data-transition look-ahead, and clock control logic that enable the clock pulse to propagate inside the latch. The circuits in the figure that are enclosed with dashed lines show the overhead associated with the internal clock gating. The functionality of each of these blocks is detailed below.

The DTLA logic compares the new input, D, with the existing output, Q. It essentially performs an XNOR function on D and Q. When $D = Q$, the

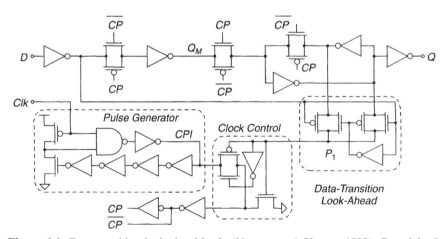

Figure 6.6. Data-transition look-ahead latch. (Nogawa and Ohtomo 1998), Copyright © 1998 IEEE.

DTLA circuit produces a logic 1 at its output, P_1, thus disabling generation of the internal clock $\{CP, \overline{CP}\}$. When $D \neq Q$, P_1 evaluates low and the clock-control (CC) circuit enables generation of $\{CP, \overline{CP}\}$, that is, the global clock, CPI, is allowed to propagate inside the latch.

The pulse-generator (PG) circuit generates a pulse, CPI, at every rising edge of the external clock, Clk. The CPI signal then triggers the latch if $D \neq Q$. The pulse generator is essential for the operation of this circuit. If there were no pulse generator, this latch could be triggered by data instead of the clock. For instance, if $D \neq Q$ and the CPI high arrives, then clock pulse, CP, is generated and Q changes. However, if D changes again while the clock is still high in such a way as to become different from Q, this also would generate pulsed pulse CP, and the latch would actually be triggered by the data. This is prevented if the Clk width is approximately equal to the CP pulse width, because then the master would be opaque while Clk is high. The pulse generator in this circuit is shared by a register of latches in order to reduce energy overhead with pulse generation. The downside of this approach is the distortion of the pulse that may occur in the clock distribution to multiple latches.

Example: Conditional Capture Flip-Flop   As an example of CSE with internal clock gating, we consider the conditional capture flip-flop (CCFF) proposed by Kong et al. (2000). The CCFF is shown in Fig. 6.7. It is a positive edge-triggered differential-input differential-output flip-flop. The CCFF is similar to the modified SAFF (M-SAFF) proposed by Oklobdzija and Stojanovic (2001). However, there is no more push-pull positive feedback, since the two coupled

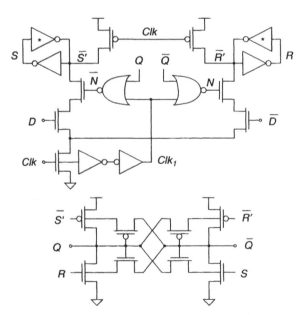

Figure 6.7. Conditional capture flip-flop. (Kong et al. 2000), Copyright © 2000 IEEE.

inverters in the M-SAFF preamplifier stage are replaced by keepers. The main difference is in the input stage where NOR gates are used to generate signals N and $\overline{N}$. These two signals regulate generation of the set and reset signals for the second-stage latch. The output stage is very similar to the output stage of the M-SAFF. The only difference is that the keepers are implemented as pass transistors instead of the full-CMOS implementation in M-SAFF.

This flip-flop operates as follows. When Clk is low, the flip-flop is in the precharge phase, $\overline{S'}$ and $\overline{R'}$ are precharged high, and the $S-R$ latch is disabled. At the rising edge of Clk, the behavior of the CCFF depends on the incoming data value — if new data are different from the previously recorded output data, one of the outputs of the NOR gates is high, enabling pull-down of $\overline{S'}$ or $\overline{R'}$. The transparency period of the differential pair is equal to the sum of two inverter delays and delay of the NOR gate. This is because N and $\overline{N}$ both go low when Clk_1 is high. During this short transparency period, new data are latched by the $S-R$ latch at the output.

It is interesting to observe that, in general, the conditional capture flip-flops have the logic function of the $J-K$ flip-flop. The CCFF is actually a $J-K$ flip-flop internally, where outputs Q and $\overline{Q}$ condition the flip-flop inputs. Inputs J and K are defined as

$$\overline{S'} = \overline{D \cdot Clk \cdot \overline{Q + Clk_1}} = \overline{D \cdot \overline{Q} \cdot (Clk \cdot \overline{Clk}_1)} = \overline{J \cdot (Clk \cdot \overline{Clk}_1)} \quad (6.3)$$

$$\overline{R'} = \overline{\overline{D} \cdot Clk \cdot \overline{Q + Clk_1}} = \overline{\overline{D} \cdot Q \cdot (Clk \cdot \overline{Clk}_1)} = \overline{K \cdot (Clk \cdot \overline{Clk}_1)} \quad (6.4)$$

where the term $Clk \cdot \overline{Clk}_1$ defines the window of time during which inputs J and K are captured. The switching activity of the internal nodes is reduced by conditioning the inputs.

6.3. DUAL-EDGE TRIGGERING

A dual-edge-triggered clocked storage element is a storage element that captures the value of the input at both clock edges. The reason for using the DETSE is to save energy in clock generation and distribution by halving the clock frequency while achieving the same throughput. Considering the increasing trends in clock frequency and clock-related energy consumption, the choice of the DETSE appears a viable method for energy reduction in the clocking subsystem.

One important consideration in the design of DETSEs is that these devices are more sensitive to the timing of the clock signal than are the single-edge-triggered clocked storage elements (SETSEs). In particular, the uncertainty of the duty cycle and the uncertainty of both clock edges become the most important design parameters. Additionally, the fact that the DETSE is more complex than the SETSE may result in longer delays and higher energy than in the corresponding SETSE.

The DETSE can be built using several techniques. Depending on the technique, we classify DETSEs as a latch-mux (LM), a pulsed-latch (PL), or a flip-flop (FF).

These classes of DETSE exhibit distinctive behavior. As discussed in Chapter 4, DETSEs have the same basic timing parameters as single-edge-triggered designs (setup time, hold time, clock-to-output delay), but applied to both clock edges. Generally, the basic timing parameters are not the same for the opposite edges, since they may be the result of different capturing mechanisms and/or different input-to-output critical paths. In the following sections, we describe in more detail the principles on which each of the techniques for building a DETSE is based. In the end, in an example of a clocking subsystem, we discuss the potential energy savings using DETSE.

6.3.1. Latch-Mux Design

The latch-mux structure is shown in Fig. 6.8. It consists of two latches connected in parallel that are transparent on opposite levels of the clock, and a multiplexer (mux) that selects the output of the nontransparent latch at all times. This structure is equivalent to a typical MSL design, but has two master latches working in parallel, and a mux functioning as a slave latch. Any MSL is therefore transferable to the corresponding latch-mux topology.

Example: Dual-Edge-Triggered Latch-Mux The dual-edge-triggered latch-mux (DET-LM) proposed by Llopis and Sachdev (1996) is shown in Fig. 6.9 as an example of the latch-mux design. It is the dual-edge counterpart of the widely used single-edge MSL proposed by Gerosa et al. (1994). The basic building blocks (latches and a multiplexer) can easily be identified on the schematic. The latches are implemented with transmission gates and clocked feedback. The multiplexer is also implemented with transmission gates. This latch-mux has two equally critical paths, somewhat shorter than the critical path of the MSL (the delay of a multiplexer versus the delay of a latch in the second stage).

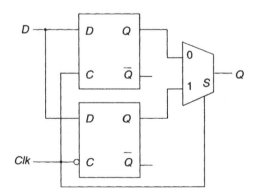

Figure 6.8. Dual-edge-triggered latch-mux design.

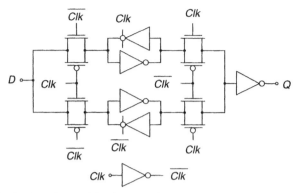

Figure 6.9. Dual-edge-triggered latch-mux circuit. (Llopis and Sachdev 1996), Copyright © 1996 IEEE.

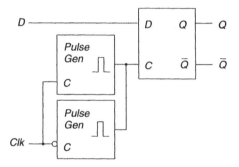

Figure 6.10. Dual-edge-triggered pulsed-latch design.

6.3.2. Pulsed-Latch Design

A conceptual diagram of a PL design is given in Fig. 6.10. It consists of a pulse generator that produces a short pulse on every edge of the clock (both leading and trailing) and a *D*-latch that is transparent for the duration of the pulse and opaque otherwise. Practical designs usually employ two pulse generators, one for each clock edge, and combine them in front of the output latch, as shown in Fig. 6.10.

Example: Dual-Edge-Triggered Pulsed Latch As an illustration of a PL design, consider dual-edge-triggered pulsed-latch (DET-PL) of Fig. 6.11. It consists of a set of input pass-gates that define the transparency window, buffer inverters, and keepers in the feedback path that keep the value stored in the PL when the latch is opaque. The transparency window is defined by the clock delay line of the four inverters. There are two timing windows when the latch is transparent — one determined by the overlap of the clock (*Clk*) and the clock delayed by the three inverters ($\overline{Clk_1}$), with another one determine by the first and fourth delay of the clock (signals $\overline{Clk}$ and Clk_2). This design is derived from

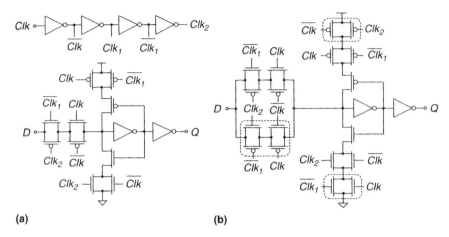

Figure 6.11. Pulsed-latch: (a) single-edge-triggered; (b) dual-edge-triggered.

the corresponding single-edge design by simply adding a transmission-gate that enables latch triggering at the trailing clock edge. Extra transmission gates in the feedback path control the keepers.

This structure does not strictly follow pulse generation and latching. The pulse generator is implicit (local to the latch), and the generated pulse is used to trigger the transmission-gate-based latch. However, the functionality of a PL still exists in terms of the pulse generation synchronously, with the clock and latching in the second stage.

6.3.3. Flip-Flop

A conceptual diagram of a DET flip-flop design is given in Fig. 6.12. It consists of two pulse-generating latches and a capturing latch (CL). The top latch creates a pulse at the leading edge of the clock, and the bottom latch creates pulse at the trailing edge of the clock *Clk*. The pulses are conditioned on data D. The CL is a nonclocked latch that captures pulses generated by the pulse-generating latches and stores the result at outputs Q, $\overline{Q}$.

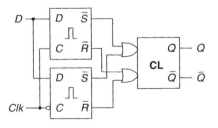

Figure 6.12. Dual-edge-triggered flip-flop design.

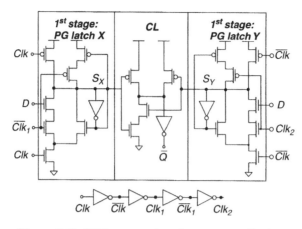

Figure 6.13. DET symmetric pulse-generator flip-flop.

Example: DET Symmetric Pulse-Generator Flip-Flop   The DET symmetric pulse-generator flip-flop (SPGFF) proposed in Nedovic et al. (2002) is shown in Fig. 6.13 as an example of DET flip-flop design. The circuit has a narrow, transparent data window and clockless output multiplexing scheme. The first stage is symmetric, consisting of two pulse-generating (PG) latches. This stage creates the data-conditioned clock pulse on each edge of the clock. The clock pulse is created at node S_X on the leading and node S_Y on the trailing edge of the clock. The second stage is a two-input NAND gate. It effectively serves as a multiplexer and a latch, implicitly relying on the fact that nodes S_X and S_Y alternate between being precharged high while the clock is low and high, respectively. This type of output multiplexing is very convenient because it does not require clock control. The clock energy is mainly dissipated by pulse generation in the first stage.

6.3.4. Clock Distribution

Both the clock distribution and CSEs have to be considered when the overall energy benefit of dual-edge-based versus single-edge-based clocking is evaluated. The best way to illustrate this is to study power savings in the clock distribution network of a single- versus dual-edge-triggered system. Generally, the dual-edge-triggered design is always a better choice than the single-edge-triggered design if its input CSE clock load is less than roughly twice that of the latter design. In addition, in systems with a significant potential wire load saving from the dual-edge-triggered scheme are even larger. This is illustrated in the following example.

The example is adapted from Nedovic et al. (2002). The crucial parameter for comparison is the total switching load, due to the storage elements, clock buffers, and wires, for the single- and dual-edge-triggered system. In this example, we find this load by estimating the load of an H-tree clock distribution network with L levels, in a microprocessor die of size $(s \times s)$ with M storage elements, as

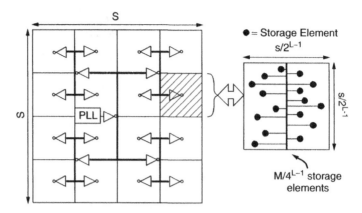

Figure 6.14. H-tree clock distribution network.

shown in Fig. 6.14. Each level-L driver supplies the clock to an area of $s/4^{L-1}$ (local domain) containing $M/4^{L-1}$ storage elements. In the local domain, the clock is distributed as shown in the shaded region of Fig. 6.14, where c_W and $C_{Clk\text{-}CSE}$ are the wire capacitance per unit length and clock capacitance of a storage element, respectively. We neglect the wire resistance, so that the width, and thus capacitance, of the wires do not depend on the storage-element clock load. Under these assumptions, it can be shown that the total load in the H-tree, including the clock load of the storage elements, is

$$C_H = \frac{4^L - 1}{3 \cdot 4^{L-1}} M C_{Clk\text{-}CSE} + c_W s \left[\frac{M}{3 \cdot 2^{L-1}} \left(\frac{4^L - 1}{4^L} \right) + \frac{1}{3} \cdot \frac{5 \cdot 4^{L-1} + 1}{2^{L-2}} - 3 \right]$$
(6.5)

The first item on the right-hand side of this equation is $C_{H\text{-}CSE}$, the portion of the H-tree capacitance that depends on the clock load of a storage element. It can be approximated to $4 \cdot M \cdot C_{Clk\text{-}CSE}/3$ if $4^L \gg 1$. The second item in the same expression is $C_{H\text{-}Wire}$, the total wire capacitance. This part of the clock distribution load is dependent only on the geometry of the H-tree, not the CSE clock load. If $M \gg 4^L \gg 1$, then $C_{H\text{-}Wire} \approx 4 c_W \cdot s \cdot M/(3 \cdot 2^{L-1})$, and the total capacitance of the H-tree becomes

$$C_H = \frac{4}{3} M \left(C_{Clk\text{-}CSE} + \frac{c_W s}{2^{L+1}} \right)$$
(6.6)

To estimate the power savings of dual-edge clocking, we assume that the number of the clock buffer levels in the H-tree is the same in dual-edge and single-edge systems. In the real design, the optimal number of levels depends on the CSE clock load, and it can be different in the two cases. However, in practical cases, the wire load dominates the CSE clock load. As a result, the optimal number of buffers is mostly affected by the wire load, rather than by the CSE clock load. Thus, the assumption of an invariant number of clock buffer levels provides a good approximation while simplifying the analysis. In addition, comparison with the same number of buffers provides approximately the same insertion delay and clock uncertainty of the clock distribution network. We define the coefficient, α, as the ratio of the clock distribution switching power consumption of the dual- and single-edge-triggered systems, assuming that dual-edge-triggered systems run at half the clock frequency of single-edge-triggered systems:

$$\alpha = \frac{P_{\text{DET}}}{P_{\text{SET}}} = \frac{1}{2}\frac{(C_{H\text{-Wire}} + C_{H\text{-CSE,DET}})}{(C_{H\text{-Wire}} + C_{H\text{-CSE,SET}})} = \frac{1}{2}\frac{(1 + C_{Clk\text{-}CSE,DET}/C_{\text{Wire-}L})}{(1 + C_{Clk\text{-}CSE,SET}/C_{\text{Wire-}L})} \quad (6.7)$$

where $C_{\text{Wire-}L} = C_W \cdot s/2^{L+1}$ is the average capacitance of the wire needed to route the clock signal from the level L buffer to a storage element, and indices DET and SET correspond to dual-edge- and single-edge-triggered clocking, respectively.

In a typical design in today's technologies, $C_{H\text{-Wire}}(W)$ is usually much larger than $C_{H\text{-}CSE}$ ($C_{Clk\text{-}CSE}$), in which case the power savings are nearly 50% greater. For example, in a five-level H-tree on a 12×12-mm die fabricated in 0.11-μm CMOS technology, the clock distribution power saving obtained by replacing a single- by a dual-edge-triggered storage element is around 40%. The plot illustrating relative power savings in the clock subsystem of the dual- versus single-edge-triggered CSE with respect to the ratio of the CSE clock capacitance to wire capacitance is shown in Fig. 6.15. The clocking power includes clock distribution, wire load, and CSEs. The savings are shown for different ratios of the clock capacitance in DET-CSE and SET-CSE.

The curves indicate that the dual-edge-triggered design is always a better choice if it maintains a clock load capacitance less than roughly twice that of the single-edge-triggered design. It should be noted that in most high-performance designs, wire load is even more pronounced than in the H-tree, design, for example, in a clock grid, so there is a larger potential saving from the dual-edge-triggered scheme.

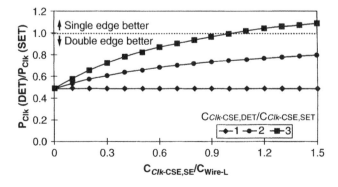

Figure 6.15. Clocking power in single- and dual-edge-triggered systems.

6.4. GLITCH ROBUST DESIGN

An interesting observation is that the best CSEs in terms of energy efficiency and large internal race margins are also the most susceptible to propagating glitches. For example, for very low input data-transition probabilities (typically less than 0.1) and relatively high glitching probabilities (greater than 0.1), the energy glitching component in the clock-gated transmission-gate MSL can become equal or even greater than the switching component. In the different CSE topologies that cover a wide range of energy consumption, rankings in energy consumption due to glitches are exactly the opposite from the rankings in spurious-free energy consumption. Specifically, conventional MSLs exhibit lower switching energy consumption than do pulse-triggered designs. However, pulse-triggered designs are less prone to glitches, which ultimately affect the robustness of the design. It is therefore important to consider the possible degradation in signal integrity because of glitches.

CSE glitch sensitivity depends on its structure. In general, the flip-flops (SDFF, HLFF, M-SAFF) exhibit greater glitch immunity than do M–S latches (MSL, C²MOS) (see Fig. 6.16). This is because internal nodes in the flip-flops are coupled with D input only during the narrow period when a flip-flop samples input data, whereas in MSLs, the master latch, when transparent, is sensitive to glitches during the whole transparency window. Circuits with internal clock gating are susceptible to glitches the most, because the glitches affect both their internal nodes and the nodes inside the clock-gating logic.

As an illustration of glitching energy consumption, Fig. 6.16 contains a comparison of the average glitching energy in various CSE topologies. In this

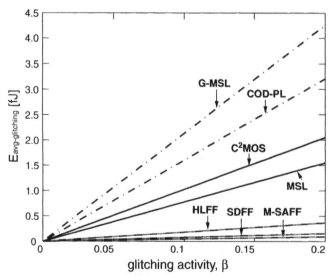

Figure 6.16. Comparison of average glitching energy in CSEs. (Markovic et al. 2001), Copyright © 2001 IEEE.

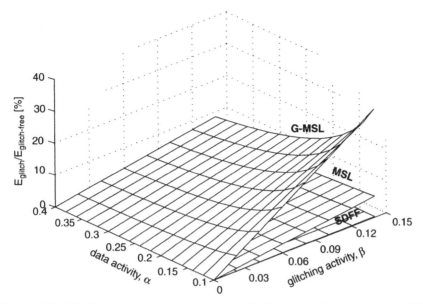

Figure 6.17. Glitching energy as a percentage of switching energy in representative CSEs showing the greatest glitch sensitivity of the gated designs.

example, it is assumed that each of the four glitches discussed in Chapter 3 occurs with equal probability: $\beta_1 = \beta_2 = \beta_3 = \beta_4 = \beta/4$. Flip-flop circuits (SDFF, HLFF, M-SAFF) consume the smallest input glitch energy because of their narrow sampling time. M–S latches (MSL, C^2MOS) are more susceptible to glitches, particularly during the half-period when the master stage is transparent. The highest glitch energy consumption of the gated designs (COD-PL, G-MSL) is due to the fact that the clock-gating logic continuously compares D and Q and propagated glitches regardless of the clock level.

Figure 6.17 shows glitching energy in a representative flip-flop, MSL, and clock-gated latch, relative to their switching energy. The figure indicates that the glitch energy is the smallest portion of the useful (glitch-free) energy in flip-flops (SDFF example), has more impact on M–S latches (MSL example), and is the most significant in designs with internal clock gating (G-MSL example).

CHAPTER 7

SIMULATION TECHNIQUES

Results and conclusions about the performance of different CSEs depend significantly on the simulation setup and evaluation environment. CSE is just one of the elements in the pipeline, and has to be sized in such a way that the optimum performance for a given output load is achieved. The CSE output loads vary a lot across the processor core, depending on the level of parallelism in each unit and also on whether the CSE is on the critical path or not.

In modern data paths CSEs experience a heavy load due to the parallel execution units and increase in interconnect capacitance. It is the performance of these CSEs on the critical path that has the highest impact on the choice of processor cycle time. Hence, in high-speed designs, the design and evaluation of CSEs is focused on the elements on the critical path and often implicitly assumes such conditions during performance comparisons. On the other hand, there are a lot of CSEs that are placed on noncritical paths with relatively light loads. While these CSEs do not directly impact the performance of the processor, careful design of these elements can significantly reduce energy consumption and alleviate clock distribution problems.

The purpose of this chapter is to recommend simulation techniques that designers can use to evaluate the performance of CSEs, depending on the desired application. Most importantly, we try to build an understanding of the issues involved in creating a simulation environment for the CSE, such that the reader can use the information to tailor his or her own setups to the specific application. There is no universal setup that is good for every CSE application.

7.1. THE METHOD OF LOGICAL EFFORT

The method of logical effort is an easy and intuitive approach to the gate-sizing problem (Sutherland and Sproull 1991). This method is especially useful in getting the initial design point right and helps to build intuition about the performance of different circuit topologies. In this section we introduce the methodology of logical-effort and describe how it can be applied to CSE sizing optimization. The logical effort methodology and reasoning are used heavily throughout this chapter, so we start with an explanation of the basic principles that will help us define the proper simulation setup and CSE performance evaluation later.

The logical-effort approach is based on an equivalent *RC* circuit model. The *RC* delay model describes delays caused by the capacitive load that the logic gate drives and by the topology of the logic gate. Inverters, as simplest logic gates, drive the loads most efficiently. More complex logic gates often require more transistors, some of which are connected in series, making them poorer drivers as compared to inverters. For example, a NAND gate introduces more delay than an inverter with similar transistor sizes, while driving the same load. The method of logical effort quantifies these effects in order to simplify delay analysis for individual logic gates and complex multistage logic networks.

The delay of a logic gate has two components: (1) a fixed component called the *parasitic delay*, p, and (2) a component that is proportional to the gate's output load called the *effort delay* or *stage effort*, f. The total delay defined in Eq. (7.1) and measured in units of technology-dependent time constant, is the sum of the effort delay and parasitic delay:

$$d = f + p \qquad (7.1)$$

The effort delay is a product of *logical* and *electrical* efforts, Eq. 7.2, which depend on the load and on the properties of the logic gate driving the load:

$$f = g \cdot h \qquad (7.2)$$

The logical effort, g, describes the effect of the logic gate's topology on its ability to produce output current. It is independent of the size of the transistors in the circuit. The electrical effort, h, characterizes the load and describes how the size of the transistors in a gate affects its driving capability:

$$h = \frac{C_{out}}{C_{in}} \qquad (7.3)$$

where C_{out} is the total output load capacitance, and C_{in} is the input gate capacitance, implicitly representing the size of the gate.

7.1.1. Multistage Logic Networks

In multistage logic networks, the method of logical effort computes the optimal number of stages and the minimum overall delay by balancing the delay among the stages. The notions of logical and electrical effort can be generalized from individual gates to multistage paths.

The logical effort, G, along a path is the product of the logical efforts of all the logic gates along the path. The electrical effort, H, of a multistage logic network is the ratio of the load capacitance at the last stage in the path to the input capacitance of the first logic gate in the path. A new kind of effort, named *branching effort*, is introduced to account for fan-out within a logic network. The branching effort, b, at the output of a logic gate is defined as

$$b = \frac{C_{on\text{-}path} + C_{off\text{-}path}}{C_{on\text{-}path}} = \frac{C_{total}}{C_{useful}} \tag{7.4}$$

where $C_{on\text{-}path}$ is the load capacitance along the analyzed path and $C_{off\text{-}path}$ is the capacitance of connections that lead off that path. If the path does not branch, the branching effort is equal to one. The branching effort, B, along the entire path is the product of the branching efforts of all the gates along the path.

As with the stage effort of individual logic gates, the *path effort*, F, is defined in the multistage logic networks as the product of the logical, electrical, and branching efforts:

$$F = G \cdot B \cdot H \tag{7.5}$$

Minimum delay along an N-stage logic network is achieved when each of the stages in the path bears the same stage effort. The minimum delay is achieved with the stage effort:

$$\hat{f} = g_i \cdot h_i = F^{1/N}, \qquad \forall i = 1, \ldots, N \tag{7.6}$$

where subscript i denotes the ith stage on the path. The minimum path delay is then

$$\hat{D} = N \cdot F^{1/N} + P \tag{7.7}$$

where P accounts for the total parasitic delay along the path. If $N = 1$, this equation reduces to Eq. (7.1).

All stages have the same effort delay, from Eq. (7.6), so once the effort delay is determined, the transistors in each logic stage are sized accordingly. Starting at the end of the path and working backwards, the input capacitance, $C_{in}(i)$, of each logic gate is determined from the capacitance transformation:

$$C_{in}(i) = \frac{g_i \cdot C_{out}(i)}{\hat{f}} \tag{7.8}$$

and appropriately distributed among the transistors in the gate connected to the input.

7.1.2. Logical Effort of Logic Gates Commonly Found in CSEs

By definition, a *static inverter* has the logical effort of one. The logical effort of other gates depends on their current-driving ability, with respect to that of an inverter, for the same input capacitance. Correspondingly, logical effort can be stated as the ratio of the gate to inverter input capacitance when both the gate and the inverter are sized so that they have the same drive current. A two-input static NAND gate with the same drive characteristics as the inverter in Fig. 7.1a is shown in Fig. 7.1b. Since the two pull-down transistors are in series, each must have twice the conductance of the inverter pull-down transistor. The input capacitance of such a static gate is 4/3 times bigger than that of the inverter with the same current-driving capability. This ratio exactly represents the logical effort of the static NAND gate with respect to the static inverter, which serves as a reference. Similarly, the pull-up transistors of a two-input static NOR gate must have twice the conductance of an inverter pull-up transistor, as shown in Fig. 7.1c. Hence the input capacitance of the static NOR gate is 5/3 times larger than that of the reference inverter for the same current drive. By analyzing the static-gate topologies, one concludes that any type of static logic gate will have a greater logical effort than the reference inverter. This, however, is not true for other circuit styles when compared to static CMOS. For example, a domino-style inverter has 2/3 times the capacitance of the static inverter for the same current drive (Fig. 7.1d). The logical effort of the domino-style inverter is 2/3 when compared to a static inverter. In the example derivations in Fig. 7.1, logical effort was calculated based on RC pull-down delays relative to that of a static inverter. In general, each input of the gate has a pull-down and pull-up logical effort, and the two are equal only if the pull-down and pull-up paths are balanced, that is, have the same current drive. Gates with unequal pull-down and pull-up logical efforts are often used to improve the performance in circuits where the logical function of the gate allows this technique to be used, that is, where one logic value is needed sooner than the other one. These gates are called "skewed" gates. An example of a skewed gate is the output inverter of a domino gate where the precharged input falling to the output of the rising inverter is the time-critical transition, requiring a good pull-up inverter drive. The pull-down inverter path is not so important, so the size of the n-MOS transistor can be decreased, which leads to a smaller total input capacitance with a constant p-MOS transistor current drive. In this way, the logical effort of the pull-up path decreased at the expense of the logical effort of the pull-down path.

Besides regular gates, some other structures, such as transmission gates or pass-transistor logic, are frequently encountered in CSE topologies. These types of logic are much harder to analyze in the context of logical effort, since their delay depends on the structure driving such a gate. The logical-effort methodology assumes that a gate is isolated from the preceding gate by its input capacitance. One easy way to include the pass gate into the logical-effort framework is to treat the pass gate and its driving gate as one complex circuit. In this context, the pass gate increases the logical effort of the resulting complex gate with respect to the driving gate alone. This occurs because of the additional series resistance of the pass gate added to the signal path. In CSE circuits it is often the case that the clock and its complement control the pass-gate transistors, as in Fig. 7.2. The resulting complex gate has two inputs, the data, A, and the

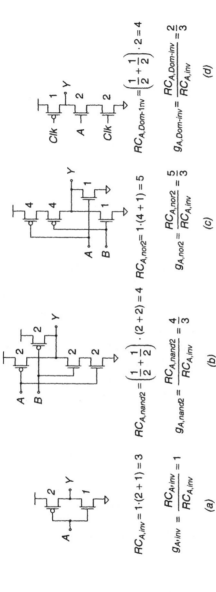

Figure 7.1. Simple logic gates: (a) reference static inverter, (b) two-input static NAND gate, (c) two-input static NOR gate, (d) domino-style inverter.

$$RC_{A,inv} = 1 \cdot (2 + 1) = 3$$

$$g_{A,inv} = \frac{RC_{A,inv}}{RC_{A,inv}} = 1$$

(a)

$$RC_{A,nand2} = \left(\frac{1}{2} + \frac{1}{2}\right) \cdot (2 + 2) = 4$$

$$g_{A,nand2} = \frac{RC_{A,nand2}}{RC_{A,inv}} = \frac{4}{3}$$

(b)

$$RC_{A,nor2} = 1 \cdot (4 + 1) = 5$$

$$g_{A,nor2} = \frac{RC_{A,nor2}}{RC_{A,inv}} = \frac{5}{3}$$

(c)

$$RC_{A,Dom\text{-}1nv} = \left(\frac{1}{2} + \frac{1}{2}\right) \cdot 2 = 4$$

$$g_{A,Dom\text{-}inv} = \frac{RC_{A,Dom\text{-}inv}}{RC_{A,inv}} = \frac{2}{3}$$

(d)

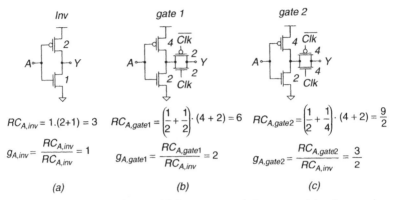

Figure 7.2. Logical effort of gate driving a transmission gate: (a) reference inverter, (b) slow-data input, (c) fast-data input sizing.

clock, Clk and $\overline{Clk}$. Depending on the transistor sizing, each input can have a different logical effort. For example, in Fig. 7.2b, transistors are sized in such a way that half the resistance on the current path is due to inverter transistors and half is due to the pass-gate transistors. In this arrangement input A has a logical effort of 2, while the logical effort of Clk input is 2/3, when compared to the comparable inverter in Fig. 7.2a. It is possible to speed up the data input by increasing the size of the pass-gate transistors, as shown in Fig. 7.2c, where logical effort is 3/2 for data input and 4/3 for Clk input. The increase in the size of the pass-gate transistors results in a delay decrease on the data path, but quickly reaches the point of diminishing returns. In addition to that, the parasitic capacitance at the output node increases, making this technique effective only in the cases when the load is much bigger than the parasitic capacitance of the transmission gate. Despite the limitations just described, this trade-off technique is heavily used in CSEs that are placed in critical paths. There, the extra clock power is often traded for a decrease in the CSE delay.

Although the logical effort is a useful tool, some modifications are needed in order to use it efficiently in the design of real submicron circuits. To simplify matters, the preceding analysis was based on a long-channel MOSFET model, suitable for back-of-the-envelope calculations. In reality, the logical effort of stacked devices is lower because of the short-channel (velocity saturation) effect, and is usually extracted from simulations. The RC delay model also fails to capture the effects of variable signal slopes on delay. However, the signal slopes tend to be equal in well-designed circuits with equal-effort delay.

7.2. ENVIRONMENT SETUP

Setting up the simulation environment is the key task of every performance comparison. The simulation setup has to be organized so that it provides the conditions for a fair comparison of different structures, yet addresses their intended application.

Several recent studies used somewhat different simulation setups, addressing different aspects of the CSE applications. We will describe some of the important concepts that these studies have addressed and present a global simulation setup framework that can be fine-tuned further for the particular application intended. The environment setup for comparing the CSEs in Stojanovic and Oklobdzija (1999) used a single-size load, chosen in a way that resembles the typical situation in a moderately to heavily loaded critical path in a processor with lots of parallelism. All the CSEs were sized so as to achieve optimum data-to-output delay for the given output load while driven from the fixed-size inverters. In most practical situations, the CSEs are designed in a discrete set of sizes, each optimized for a particular load. Hence, it is very useful to examine the performance of the CSE for a range of loads around the load for which the CSE was optimized. This technique is illustrated in Nikolic and Oklobdzija (1999), where different CSEs are initially sized to drive a fixed load, and the load is then varied. In this setup, the delay of a CSE will exhibit linear dependence on the load, with the slope of the delay curve illustrating the logical effort of the driving stage of the CSE, and the zero-load crossing illustrating the parasitic delay of the driving stage together with the delay of the inner stages of the CSE. As we will see in the remainder of this section, this is not the optimal behavior of the CSE delay curve, but is the best that can be achieved when there are only a few CSE sizes available in the library. In the case where the CSE can be reoptimized for each particular load, further speedup can be achieved, since the effort can be shared between stages rather than relying solely on the output stage. This approach was illustrated by Heo and Asanovic (2001). However, contrary to the conclusions in that paper, in the case where the general performance of a CSE needs to be assessed, the proper approach is to optimize the CSE for the most important application that determines the performance of the whole system, not the most frequent application. In high-speed systems, the most important are the elements on the critical path, which is typically moderately to heavily loaded due to branching to parallel execution units and wire capacitance. The small number of critical paths in a processor does not decrease their importance, since it is their delay that determines the clock rate of the whole system. The performance of a large number of lightly loaded CSEs that are placed off the critical path is of concern only if it can be traded for energy savings.

The simulation approach should attempt to approximate the actual data-path environment. The number of logic stages in a CSE and their complexity are very dependent on a particular circuit implementation, which leads to differences in logical effort, parasitic delay, and energy consumption. Every CSE structure needs to be optimized to drive the load with the best possible effort delay.

A general simulation setup is illustrated in Fig. 7.3. The size of the data input is fixed for all CSEs in order to exclude the impact of pipeline logic on the CSE comparison. The data signal has a signal slope identical to that of an FO4 inverter, which is the case in a well-designed pipeline. This setting is typical in designs where delay and energy requirements are balanced. On the other hand, in the high-speed design methodology of Intel, Sun Microsystems, and the former

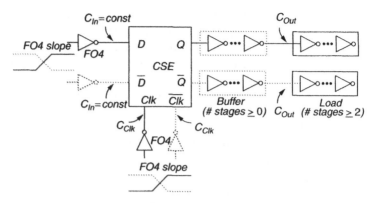

Figure 7.3. General simulation setup.

Digital Corporation, FO3 inverter metric is more common than FO4 because of a more aggressive design style.

The size of the clocked transistors is set to the size needed in order not to compromise the speed of the whole structure. As discussed in the previous section, a direct trade-off exists between the CSE delay and clock energy (size of clocked transistors), as some of the clocked transistors are always on the critical path of the CSE. An increase in the sizes of the clocked transistors on a critical path results in diminishing returns, since data input is fixed. Depending on the CSE topology, some structures can trade delay for clocked transistor size more efficiently than others, so we allow this to happen up to a certain point. Our goal here is to examine CSEs that are used on a critical path, hence the assumption that the designer might be willing to spend a bit more clock power to achieve better performance. Differences in clock loads (C_{Clk}) among devices illustrate potential drawbacks in terms of clock power requirements, and serve as one of the performance metrics. Clock inputs have a signal slope that is identical to that of an FO4 inverter. This can be changed depending on the clock distribution design methodology.

The question of how to compare differential and single-ended structures has always been one of the key issues among the people characterizing and designing CSEs. The immediate answer, the most fair, and at the same time the easiest one, is that differential and single-ended structures should not be compared with each other, due to the cost that single-ended structures incur in generating the complementary output. We have decided to follow the other approach, and not require that single-ended structures generate both true and complementary values at the output.

The worst-case analysis requires that the CSE generate the output that has worst data-to-output delay. However, it is also beneficial to measure both the $D-Q$ and $D-\overline{Q}$ delay. Any imbalance between the two can lead to big delay savings in cases where proper logic polarity manipulation in the stages preceding or following the CSE can change the polarity requirement of the CSE, and hence its data-to-output delay. The load model always consists of several inverters in a chain to avoid the error in delay caused by Miller capacitance effects from the fast switching load back to the driver.

The logical-effort framework offers analogy between the CSE and a simple logic gate. At light load, the logic gate is dominated by its parasitic delay, that is, self-loading. At high load, the effort delay becomes the dominant factor. Similarly, at light load, delay of a CSE with large number of stages is entirely dominated by parasitic delay. However, at high load, more stages are beneficial in reducing the effort delay, which then dominates over parasitic delay. Therefore, the performance of the CSE is best assessed if it is evaluated in a range of output loads of interest for the particular application.

CSE evaluation can either be performed using some representative critical-path load or a set of loads can be used, in which case the CSE has to be reoptimized for each load setting. Depending on the choice of the output load size, some CSE structures with an inherently small number of stages and high logical effort may require additional buffering in order to achieve the best-effort delay. This is shown in Fig. 7.4, where for some fixed C_{in}, the output load C_{out} is optimally driven by the CSE with logical effort G_{CSE} and K stages, and additional $N-K$ levels of inverters.

Now, for each CSE we need to find the optimal effort per stage and number of stages to drive the required load, as shown in Eqs. (7.7)–(7.9)

$$H = \frac{C_{out}}{C_{in}} \tag{7.7}$$

$$N = \text{round}(\log_4(G_{CSE}H)) \tag{7.8}$$

$$f = \sqrt[N]{G_{CSE}H} \tag{7.9}$$

Starting from total electrical fan-out, H, the optimal number of stages, N, is obtained by rounding the logarithm of the total path effort (assuming g_{Inv} is unity). The logarithm is of base 4, since a stage effort of 4 is a target for optimal speed. Once the integer number of stages is obtained, an updated value

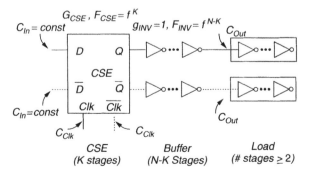

Figure 7.4. Additional buffering in simulation test bed.

of the stage effort is found from Eq. (7.9). After the stage effort is obtained, CSE internal stages have to be resized for the new stage effort as well as for the external inverters, if there are any.

This sizing approach is optimal even in the case where no additional inverters are required, since it will serve to distribute the effort between the internal stages of the CSE. We now illustrate the CSE sizing in examples of two widely used flip-flops.

7.2.1. HLFF Sizing Example

In this example we observe the change in minimum data-to-output ($D-Q$ or $D-\overline{Q}$) delay as the output load of the CSE increases. Before we start investigating the effect of different loads on the sizing of the HLFF, let us show how the logical effort can be calculated for the given sizing, as shown in Fig. 7.5. It is relatively easy to see that the HLFF is made up of a three-input static NAND gate as the first stage and a domino-like three-input NAND gate in the second stage. Minor variations from standard static NAND sizing for equal logical effort on all inputs are needed to speed up the data input and enable the first stage to evaluate before the transparent window closes. There is a similar situation in the second stage. This HLFF sizing example also illustrates the application of logical effort to skewed gates (gates in which one output transition is faster than the other) and gates with keepers.

The critical path of the HLFF is exercised with a 0-to-1 transition at data input. The first stage of the HLFF is a skewed NAND gate. This is because one

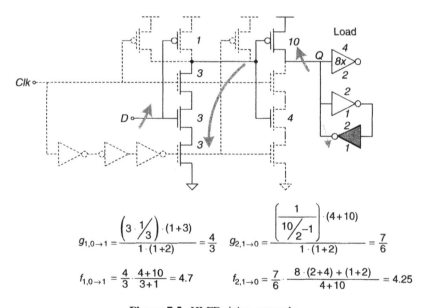

$$g_{1,0 \to 1} = \frac{\left(3 \cdot \dfrac{1}{3}\right) \cdot (1+3)}{1 \cdot (1+2)} = \frac{4}{3} \qquad g_{2,1 \to 0} = \frac{\left(\dfrac{1}{10/2 - 1}\right) \cdot (4+10)}{1 \cdot (1+2)} = \frac{7}{6}$$

$$f_{1,0 \to 1} = \frac{4}{3} \cdot \frac{4+10}{3+1} = 4.7 \qquad f_{2,1 \to 0} = \frac{7}{6} \cdot \frac{8 \cdot (2+4) + (1+2)}{4+10} = 4.25$$

Figure 7.5. HLFF sizing example.

of the inputs to the NAND gate is the clock, which precharges the output of the first stage before data can go through the first stage. Thus, the role of the data-controlled p-MOS transistors is that of a keeper in case the data start changing from 1-to-0 within the transparency window, that is, after the rising edge of the clock. The logical effort and stage effort of the first stage are calculated as shown in Fig. 7.5, for a 0-to-1 transition at data input. The logical-effort calculation of the second stage is slightly more complicated because of the keeper–inverter pair. A keeper sinks a portion of the current that is sourced by the p-MOS transistor to node Q. Therefore, it can be considered as negative conductance. In Fig. 7.5, this negative conductance is found by subtracting the conductance of the n-MOS transistor (1) of the shaded keeper inverter from the conductance of the driving p-MOS transistor (10/2). For the particular load given in Fig. 7.4, efforts per stage were calculated to be 4.7 and 4.25, which is near the optimum value of 4, indicating that this example sizing is nearly optimal. The reader is cautioned, however, that the sizing in this example is somewhat simplified, because the short channel stack effect has not been taken into account and the logical-effort values for the n-MOS transistor stack are somewhat pessimistic. Once the logical effort of each stage is known, it can be used to adjust the sizing of each stage as the load is increased or decreased. The alternative method is to use one of the automated circuit optimizers; however, we do not recommend it as the initial method, simply because it is essential that the designer gets to know the circuit through manual sizing and logical-effort estimation. This builds intuition about the circuit and the ability to verify optimizer results.

The performance of three different sizing solutions versus the electrical effort (fanout) is given in Table 7.1, where the data-to-output delays are normalized to the FO4 inverter delay. While there is only one optimal solution for each load size, in this example we examine only three cell sizes in order to illustrate the principle in a simple manner.

When the load is relatively small, just adjusting the size of the internal stages and balancing the stage effort can achieve speedup, as with cell sizes A and B. For large loads, an additional inverter stage is needed to bring the stage effort close to four. The optimal delays are set in bold in Table 7.1 to illustrate the change in the optimal sizing selection with the increase in electrical effort. An interpolated version of these data is shown in Fig. 7.6, where all three sizing cases are illustrated along with the best sizing versus fanout curve. According

Table 7.1 HLFF Delay (normalized to FO4 inverter delay) vs. Fanout for Different HLFF Cell Sizes

Fan-out	4	16	42	64	128
Load size (# stages)					
Small-A (2)	**1.60**	**2.06**	3.11	4.19	7.80
Medium-B (2)	1.80	**2.06**	**2.59**	3.05	4.62
Large-C (2 + 1)	2.27	2.44	2.74	**2.96**	**3.56**

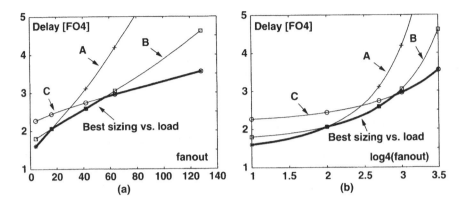

Figure 7.6. Sizing versus load, HLFF example: (a) linear, (b) log4 scale.

to the logical-effort theory, the optimal delay versus fanout curve should have logarithmic shape, which indeed holds for the "best sizing vs. load curve" in Fig. 7.6a. Similarly, the optimal delay is a linear function of the logarithm of the electrical effort (fanout), as shown in Fig. 7.6b. The logarithmic fan-out scale makes it really easy to see if the stage effort is properly determined. Recall that from Eqs.(7.6) and (7.9) delay is a linear function of the stage effort and the number of stages. The logarithm of the electrical effort approximately illustrates the required number of stages, and if the delay checks out to be a multiple of the number of stages and FO4 delay, then the optimal effort per stage is chosen. This is the case in Fig. 7.6b with all the curves that are best in a certain range of loads.

Although it is interesting to evaluate the behavior of a CSE for a wider range of loads, most often it is required that the CSE operates well in a much narrower range. Typically, high-performance CSEs are placed in critical paths with a relatively high average load. Thus, in that case, a one-point performance comparison can be made for some preselected value that describes the average load of the CSE on a critical path. In this book, all the comparisons have been made using the moderately high electrical effort of 42 (third column in Table 7.1)

While the application of logical-effort analysis is applied to HLFF in a straightforward fashion due to the easily recognizable circuit topology, it can also be applied to more exotic circuits, such as M-SAFF, that contain sense amplifiers and other structures not covered in the introductory section on logical effort. It is important to note that logical effort can be calculated or simulated for any circuit topology.

7.2.2. M-SAFF Sizing Example

In this problem we also observe the minimum $D-Q$ delay as the output load of the CSE increases. The performance of three different sizing solutions is

Table 7.2 M-SAFF Delay vs. Fanout for Different M-SAFF Cell Sizes

Fanout	4	16	42	64	128
Load size (# stages)					
Small-A (2)	**2.33**	2.60	3.11	3.53	4.70
Medium-B (2)	2.35	**2.59**	**3.01**	**3.34**	4.24
Large-C (2 + 1)	3.06	3.15	3.31	3.44	**3.83**

illustrated versus the electrical effort, normalized to the delay of the FO4 inverter, in Table 7.2.

The sizing is done in a way similar to that described in the HLFF example. The only caveat with the M-SAFF structure is to recognize that the logical effort of the input stage is very small, better than that of an inverter, because of the small input capacitance. This implies that the sizing changes will mostly be located in the output stage since the input stage can accommodate larger load variations without needing to be resized. While it was relatively easy to find different sizes that perform better at certain loads in the case of HLFF, this is not so in the case of M-SAFF. The small logical effort of the whole structure enables it to cover a huge range of loads, with a single size achieving relatively good performance. This is the case with the structure of size B in Table 7.2. In Fig. 7.7 size A is only slightly better than size B, and only for a very light load of FO4; subsequently the size B device takes the lead all the way up to the FO64, after which an additional inverter is needed to prevent excessive delay.

7.2.3. Energy Measurements

While so far we have been mostly concerned about the performance aspects of the simulation setup, it is very important to prepare the simulation environment correctly, so that the energy parameters of the CSE are measured accurately.

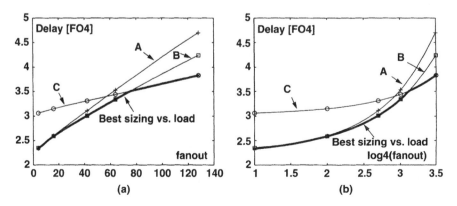

Figure 7.7. Sizing versus load, M-SAFF example: (a) linear, (b) log4 scale.

We only need to set the measurements to capture the energy for each of four possible binary transitions. Accurate average energy estimates can be made with these values, based on the statistics of the incoming data. Using state-transition diagrams (Zyban and Kogge 1999), more formal methods can be used to exactly evaluate the effect of regular transitions and glitches in the total switching energy of the CSE. In line with energy breakdown in Chapter 3, in order to measure different energies, it is essential to provide separate supply voltages for different stages of the CSE.

7.2.4. Automating the Simulations

The delay versus load CSE evaluation described in the preceding examples can be implemented automatically. Here we outline the procedure for creating such an automated simulation environment. The authors suggest *Perl* as one of the most convenient scripting languages today.

For each CSE, we need to determine the logical effort of every stage based on its topology (e.g., two NAND-like stages, one inverter stage, would be 4/3, 4/3, 1), or better yet, exact logical-effort values obtained from the simulation. A very good starting point can be the *Perl* script that characterizes logical effort, FO4 delay, and much other data for a given technology process, freely available from Sutherland et al. (1999). The product of the logical efforts of all the stages should equal the total logical effort of the CSE. After the total logical effort is found, the optimal number of stages and updated stage effort can be calculated from Eqs. (7.7)–(7.9). Now, with stage effort and logical efforts having been obtained from the topology of the CSE, taking the data input of a fixed size, and assuming that the clock is on (i.e., treating the structure as a cascade of logic gates), transistor sizes for every stage can be calculated, progressing from the data input to the final load in the simulation setup.

When a library of CSEs is created, a presimulation should be run for each environment parameter setup. This run should include various process corners and supply voltages, in order to determine the FO4 inverter slope and set that value as the rise/fall time of signals that drive the data and clock into the CSE. A simulation of the flow is given in Fig. 7.8.

For each device in the library, $D-Q(\overline{Q})$ delay and $Clk-Q(\overline{Q})$ delay are stored in each run, decreasing the delay between the edge of the input data and clock edge (setup time). The script should check for the setup/hold time failure (i.e., when the CSE fails to pass the input value to the output). This is typically detected by the long $Clk-Q$ delay (i.e., the measurement target occurred in the next cycle) or failure to measure the delay if only one cycle is simulated. The script automatically finds the minimum delay point at all the specified outputs. The whole procedure is repeated for a range of loads and the best sizing curve is found, as shown in Figs. 7.6 and 7.7. The following appendix of this book contains an example script written in *Perl* that can serve as the basis of a more sophisticated tool for CSE characterization. In addition to the script, we also provide example spice decks for HLFF and M-SAFF used in this example. These files are a good start for a designer who wants to evaluate various CSE topologies.

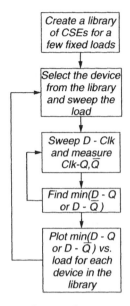

Figure 7.8. CSE flow simulation.

7.3. APPENDIX

This appendix contains a few useful scripts for characterization of the CSEs. A list of the files is given below:

1. *Perl* script for CSE characterization;
2. Parameterized spice deck for FO4 inverter delay extraction (called by the CSE characterization script);
3. Parameterized spice deck for CSE characterization;
4. Example spice circuit decks for HLFF and M-SAFF.

7.3.1. The CSE Characterization Script

```
#!/usr/local/bin/perl

##########################################################

# charCSE.pl

##########################################################

#
# The script expects you to pass the nominal operating
    voltage, value for lambda,
```

```
# and path to the hspice models library.
#
# Usage:  charCSE.pl VDD lambda processfile
#
########################################################
#
########################################################

# Load Libraries

########################################################

#
require "ctime.pl";

########################################################

# Start Script

########################################################


# Check for proper number of arguments and extract args
if ($#ARGV !=2) {#$#ARGV is number of command line
    arguments minus 1
      print STDERR "Usage: $0 VDD LAMBDA modelsfile\n";
      #$0 is script
      print STDERR "ex:\n $0 2.5 0.15u opConditions
      .lib\n";
      exit;
}


$VDD = $ARGV[0];
$LAMBDA = $ARGV[1];
$processfile = $ARGV[2];
$diffwidth =5; # lambda width of diffusion from gate


#set the min-max temperature and voltage parameters
   here or in the library call
$THIGH = 100;
$TNOM =  75;
$TLOW =   0;
$VHIGH = 1.1*$VDD;
$VLOW =  0.9*$VDD;


#Start timer
$time = time;
$a = 0;
$b = 0;
```

```
############################################################

# Run Simulations

############################################################

#Select the applicable sim or run them all
   (1=yes, 0=no).

$runall= 0;
$run1 = 0; #Inverter: Delay vs. FANOUT
$run2 = 1; #FF delay vs. FANOUT, FF resized for each
   FANOUT

############################################################

# Part (1):  Inverter delay vs. fanout (reference)

############################################################

#
@casesINV = ('Rise','Fall','Avg');
@fanoutsINV=(2,4,6,8);
if ($runall || $run1) {
   foreach $fanout (@fanoutsINV) {
       foreach $case (@casesINV) {
               $resultINV[$fanout]{$case} = &runsim
   ("simInv", "d$case","!TEMP!",$THIGH,
                            "!CORNER!", "TT", "!VOLT!",
   $VLOW, "!FANOUT!", $fanout);
           }
   }
}
############################################################
# Part (2):  FF sims vs. fanout
############################################################
# specify desired measurement parameters
#@casesDUT = ('cqRise','cqFall','cqbRise','cqbFall',
   'setupQRQbF','setupQFQbR');
#@casesDUT = ('dqr','dqbf','dqf','dqbr');
@casesDUT = ('dqr','dqf');#('dqbf','dqbr');

#specify CSE name from the library of sizes and CSE
   types
$nameDUT='hlff4';
@fanoutsDUT = (4,16,42,64,128); #values of the fanout
   sweep
$dclk_p=50e-12; $dclk_n=0e-12; #dclk_step=5e-12;
```

```perl
   #start, end and step of D-Clk sweep
$dqmax=1e-9; # d-q delay fail-check value
# initialize delays
foreach $fanout (@fanoutsDUT) {
  foreach $case (@casesDUT) {
      $delay[$fanout]{$case}=$dqmax;
    }
}

$j=0;
if ($runall || $run3) {
   foreach $fanout (@fanoutsDUT) {
       for($dclk=$dclk_p; $dclk>=$dclk_n; $dclk=$dclk
          -$dclk_step) {
             foreach $case (@casesDUT) {
               $tmpDUT[$case] = &runsim("sim","$case","
                 !TEMP!",$THIGH,
                                       "!CORNER!", "TT",
                                       "!VOLT!", $VDD,
                                "!DUT!",$nameDUT,"!FANOUT!",
                               $fanout,"!SETUP!",$dclk);
          print "$tmpDUT[$case] ";
             if ($delay[$fanout]{$case} > $tmpDUT
               [$case]) {
                   if($tmpDUT[$case] > 0){
                      $delay[$fanout]{$case} = $tmpDUT
                        [$case];
                   }
             }
             printf(STDOUT "%2.0f\n", 1e12*$delay
               [$fanout]{$case});
          }
          $j=$j+1;
       }
     }
}

###########################################################

# Finish timer

$date = &ctime(time);chop($date); # get current date
   and strip newline
$time = time-$time;

###########################################################

# Print Results

###########################################################
```

```
# This section prints out all the results.
open (PRINTOUT, ">cond.out");
print PRINTOUT "\n*** Process Characterization Results
   ***\n\n";
print PRINTOUT "Process File: $processfile \n";
print PRINTOUT "VDD: $VDD LAMBDA: $LAMBDA \n";
print PRINTOUT "Operating Conditions: $opconditions \n";
print PRINTOUT "Device lengths minimum size.\n";
print PRINTOUT "Run completed $date in $time seconds
   \n\n";
close (PRINTOUT);

###########################################################

if ($runall || $run1){
open (PRINTOUT, ">inv.out");
foreach $fanout (@fanoutsINV) {
    printf(PRINTOUT "%2.0f", $fanout);
    foreach $case (@casesINV) {
        printf(PRINTOUT " %3.2f ", $resultINV[$fanout]
           {$case}*1e12);
      }
    printf(PRINTOUT "\n");
  }
close (PRINTOUT);
}

###########################################################

#
if ($runall || $run2){
open(PRINTOUT, ">dut$nameDUT.out");
  foreach $fanout (@fanoutsDUT) {
         printf(PRINTOUT "%2.0f", $fanout);
      foreach $case (@casesDUT) {
          printf(PRINTOUT " %3.2f ", $delay[$fanout]
             {$case}*1e12);
        }
   printf(PRINTOUT "\n");
}

close (PRINTOUT);
}

###########################################################

# Subroutines

###########################################################

#
```

```
# The subroutines provide a quick way to start
  simulations and extract the results from SPICE.
#
# sub runsim
#
# This subroutine takes the name of the HSPICE deck
  and a measured
# parameter value to extract as inputs.
# It substitutes the values of VDD and processfile you
  provide
# for the variables !VDD! and !LIB! in the HSPICE
  deck and
# also handles any other substitutions you passed.
    It then
# runs HSPICE and extracts the value of the parameter
  you specified found
# by a .measure statement in the deck.

sub runsim {
  $olddeckname = $deckname;
  $oldmear = $measure;
  @old = @subs; # save old arguments
  @subs = @_; # Grab list of arguments to substitute

  $deckname = shift(@subs); # Grab deckname passed
     to runsim
  $measure = shift(@subs); # Grab parameter to measure

  print "Extracting $measure from $deckname with:\n  ";
  for ($i=0; $i<=$#subs; $i+=2) {
      print "$subs[$i] = $subs[$i+1]   ";
  }
  print "\n";

  # If old arguments are the same as new ones,
    recycle old results
  $recycle = 1;
  for ($i=0; $i<=$#subs; $i++) {
      if (($old[$i] ne $subs[$i])||($olddeckname ne
        $deckname)) {
#        ($oldmear ne $measure)) {
        $recycle = 0;
        }
}
if ($recycle != 1) {
      # Open the spice deck and a temporary output file
      open(DECK, $deckname.".hsp")
        || die("Can't open $deckname.hsp: $!\n");
      open(OUT, "> tmp_deck.hsp") || die("Can't open
        tmp_deck.hsp: $!\n");
```

```perl
    # Read each line of the deck, substitute VDD &
        procfile, write out
    while (<DECK>) {
     s/!SUP!/$VDD/g;
     s/!LAMBDA!/$LAMBDA/g;
     s/!LIB!/$processfile/g;
     for ($i=0; $i<=$#subs; $i+=2) {
         s/$subs[$i]/$subs[$i+1]/g; # replace all
            occurrences
     }
     print OUT $_;
}
print "Not recycling\n";
# Close files
close(OUT);
close(DECK);




# Run HSPICE simulation
# Close STDERR while running to avoid messages printed
           by SPICE
open(SAVEERR, ">&STDERR");
close(STDERR);
system("hspice tmp_deck.hsp > tmp_deck.out");
open(STDERR, ">&SAVEERR");
close(SAVEERR);
}
    # Extract result from output file
    open(RESULT, "tmp_deck.out") || die("Can't open
      tmp_deck.out: $!\n");
    $result = "";
    while (<RESULT>) {
        if (/\*error/) { # HSPICE produced an error
            print STDERR "$_";
            $next = <RESULT>;
            die("$next");
        }
        if (/\s*$measure\s*=\s*(\S+)/i) { # Search for
             $measure = xxx
            $result = $1; last; # and record xxx
        }
    }
    if ($result eq "") {
        die ("Couldn't find $measure\n");
    }
    return $result;
}}
```

7.3.2. Simulation Bench for FO4 Inverter Delay Extraction (simInv.hsp)

```
************************************************************

*Set supply and library

************************************************************


*The script replaces !LIB! with the user-specified
   library.
*It also sets the operating conditions and process
   corner.

.protect                   *Don't print the contents of
                                   library
.lib 'opConditions.lib' TT *Load the library for
                                   process corner
.unprotect             *Resume printing SPICE deck

.temp !TEMP!
.param Supply=!VOLT!       *Set characterization voltage
.opt scale=!LAMBDA!        *Set lambda

.param ct = 10n  *cycle time for the clock pulse source
.param rt = 0.1n  *rise time for the pulse source

*Save results of simulation for viewing
.options post

************************************************************

*Define power supply

************************************************************

.global Vdd   Gnd
Vdd       Vdd   Gnd    'Supply'         *Supply is set by
                                         .lib call

************************************************************

*Define Subcircuits

************************************************************
```

```
.SUBCKT inv in out WP=16 LP=2 WN=8 LN=2
M_0 out in Gnd Gnd NMOS W=WN L=LN GEO=1
+ AS='(WN)*6.1' AD='(WN)*3.85'
+ PS='3.00*(WN)+2*6.1' PD='2.00*(WN)+2*3.85'
+ NRD='3.85/(WN)' NRS='3.85/(WN)'
M_1 out in Vdd Vdd PMOS W=WP L=LP GEO=1
+ AS='(WP)*6.1' AD='(WP)*3.85'
+ PS='3.00*(WP)+2*6.1' PD='2.00*(WP)+2*3.85'
+ NRD='3.85/(WP)' NRS='3.85/(WP)'
.ENDS $ inv


**********************************************************


* Top level simulation netlist


**********************************************************


x1  In   Inb  inv              *set appropriate slope
x2  Inb  Inv  inv  M='!FANOUT!'      *drive real load
xl1 Inv  Dm1  inv  M='!FANOUT!*!FANOUT!' *real load
xd1 Dm1  DD1  inv  M='!FANOUT!*!FANOUT!*!FANOUT!'
                              *load on load-important

**********************************************************

* Stimulus

**********************************************************

*Format of pulse input:
*pulse v_initial v_final t_delay t_rise t_fall
   t_pulsewidth t_period

Vin In Gnd pulse 0 'Supply' 1ns 'rt' 'rt' 'ct/2-rt'
    'ct'

*Set Initial Conditions to insure no false transitions
    during
*initialization

.IC V(Inb)='Supply' V(Inv)=0 V(Dm1)='Supply' V(DD1)=0
**********************************************************

*Measurements

**********************************************************

.measure dRise
+    TRIG v(Inb) VAL='Supply/2' FALL=1
+    TARG v(Inv) VAL='Supply/2' RISE=1
.measure dFall
```

```
+   TRIG v(Inb) VAL='Supply/2' RISE=1
+   TARG v(Inv) VAL='Supply/2' FALL=1

.measure dAvg param='(dRise+dFall)/2'

.tran .001ns 12ns

**********************************************************

*End of Deck

**********************************************************

.end
```

7.3.3. CSE Simulation Bench in SPICE (sim.hsp)

```
**********************************************************

*Set supply and library

**********************************************************

*The script replaces !LIB! with the user-specified
   library.
*It also sets the operating conditions and process
   corner.


.param Sup=!SUP!

.protect                 *Don't print the contents of
                            library
.lib 'opConditions.lib' !CORNER!   *Load the library
                                    for process corner
.unprotect           *Resume printing SPICE deck
.temp !TEMP!
.param Supply=!VOLT!   *Set characterization voltage
.opt scale=!LAMBDA!     *Set lambda

.param cc = 2n  *cycle time for the clock pulse source
.param cd = '4*cc' *cycle time for the data pulse source
.param rt = 0.1n  *rise time for the pulse source

*Save results of simulation for viewing
```

```
.options post

*********************************************************

*Define power supply

*********************************************************

.global Vdd  Gnd
Vdd Vdd Gnd 'Supply' *Supply is set by .lib call

*********************************************************

*Define Subcircuits

*********************************************************
.SUBCKT inv in out WP=16 LP=2 WN=8 LN=2
M_0 out in Gnd Gnd NMOS W=WN L=LN GEO=1
+ AS='(WN)*6.1' AD='(WN)*3.85'
+ PS='3.00*(WN)+2*6.1' PD='2.00*(WN)+2*3.85'
+ NRD='3.85/(WN)' NRS='3.85/(WN)'
M_1 out in Vdd Vdd PMOS W=WP L=LP GEO=1
+ AS='(WP)*6.1' AD='(WP)*3.85'
+ PS='3.00*(WP)+2*6.1' PD='2.00*(WP)+2*3.85'
+ NRD='3.85/(WP)' NRS='3.85/(WP)'
.ENDS $ inv

.include '!DUT!.hsp'

*********************************************************

*Top level simulation netlist

*********************************************************

.param nstages=2
*xD      D    Dl     inv wp=4 wn=4 m='FO-1'
*xDdum   Dl   Dldum  inv wp=4 wn=4 m='FO*(FO-1)'
*xDb     Db   Dbl    inv wp=4 wn=4 m='FO*(FO-1)'
*xDbdum  Dbl  Dbldum inv wp=4 wn=4 m='FO*(FO-1)'

xDUT   Clk D Db Q Qb !DUT!
xQ     Q    Ql    inv wp=16 wn=8  m='!FANOUT!'
xQdum   Ql   Qldum    inv wp=16 wn=8  m='4*!FANOUT!'
*xQb     Qb   Qbl   inv wp=16 wn=8  m='!FANOUT!'
*xQbdum  Qbl  Qbldum  inv wp=16 wn=8  m='4*!FANOUT!'
```

```
***********************************************************

*Stimulus

***********************************************************

*Format of pulse input:
*pulse v_initial v_final t_delay t_rise t_fall
   t_pulsewidth t_period

Vd   D   Gnd   pulse   0 'Supply' 'cc-!SETUP!' 'rt' 'rt'
     'cd/2-rt' 'cd'
Vdb Db Gnd   pulse 'Supply' 0   'cc-!SETUP!' 'rt' 'rt'
     'cd/2-rt' 'cd'
Vclk Clk Gnd pulse 0 'Supply' 0 'rt' 'rt' 'cc/2-rt' 'cc'

*Set Initial Conditions to insure no false transitions
   during
*initialization
.IC V(xDUT.Q)=0 V(xDUT.Qb)='Supply'

***********************************************************

*Measurements

***********************************************************

.measure cqRise
+    TRIG v(Clk)   VAL='Supply/2' RISE=2
+    TARG v(Q)     VAL='Supply/2' RISE=1
.measure cqFall
+    TRIG v(Clk)   VAL='Supply/2' RISE=4
+    TARG v(Q)     VAL='Supply/2' FALL=1

.measure cqbRise
+    TRIG v(Clk)   VAL='Supply/2' RISE=4
+    TARG v(Qb)    VAL='Supply/2' RISE=1
.measure cqbFall
+    TRIG v(Clk)   VAL='Supply/2' RISE=2
+    TARG v(Qb)    VAL='Supply/2' FALL=1

.measure setupQRQbF
+    TRIG v(D)     VAL='Supply/2' RISE=1
+    TARG v(Clk)   VAL='Supply/2' RISE=2

.measure setupQFQbR
+    TRIG v(D)     VAL='Supply/2' FALL=1
```

```
+    TARG v(Clk)     VAL='Supply/2' RISE=4
.measure dqr  PARAM='cqRise+setupQRQbF'
.measure dqbf PARAM='cqbFall+setupQRQbF'
.measure dqf PARAM='cqFall+setupQFQbR'
.measure dqbr PARAM='cqbRise+setupQFQbR'


.tran .001ns '5*cc'


***********************************************************


*End of Deck


***********************************************************


.end
```

7.3.4. Example HLFF Deck (hllf16.hsp)

```
*FILE: hlff16.hsp
*SPICE netlist for "hlff"
*start main CELL hlff

.SUBCKT hlff16 Clk D Db Q Qb
Xinv145 Q Qb inv WP=8 LP=2 WN=4 LN=2
Xinv152 Qb Q inv WP=8 LP=2 WN=4 LN=2
Xinv159 Clk net_1 inv WP=8 LP=2 WN=4 LN=2
Xinv264 net_2 Clkbbb inv M=4 WP=8 LP=2 WN=4 LN=2
MnD net_3 D net_4 Gnd NMOS W=8 L=2 GEO=1 M=2
+ AS='(8)*6.1' AD='(8)*3.85'
+ PS='3.00*(8)+2*6.1' PD='2.00*(8)+2*3.85'
+ NRD='3.85/(8)' NRS='3.85/(8)'
Xinv399 net_1 net_2 inv M=4 WP=8 LP=2 WN=4 LN=2
MpClk X Clk Vdd Vdd PMOS W=4 L=2 GEO=1
+ AS='(4)*6.1' AD='(4)*3.85'
+ PS='3.00*(4)+2*6.1' PD='2.00*(4)+2*3.85'
+ NRD='3.85/(4)' NRS='3.85/(4)'
MpClkbbb X Clkbbb Vdd Vdd PMOS W=4 L=2 GEO=1
+ AS='(4)*6.1' AD='(4)*3.85'
+ PS='3.00*(4)+2*6.1' PD='2.00*(4)+2*3.85'
+ NRD='3.85/(4)' NRS='3.85/(4)'
MnClk X Clk net_3 Gnd NMOS W=8 L=2 GEO=1 M=4
+ AS='(8)*6.1' AD='(8)*3.85'
+ PS='3.00*(8)+2*6.1' PD='2.00*(8)+2*3.85'
+ NRD='3.85/(8)' NRS='3.85/(8)'
MnClkbbb net_4 Clkbbb Gnd Gnd NMOS W=8 L=2 GEO=1 M=4
+ AS='(8)*6.1' AD='(8)*3.85'
```

```
+ PS='3.00*(8)+2*6.1' PD='2.00*(8)+2*3.85'
+ NRD='3.85/(8)' NRS='3.85/(8)'
MpD X D Vdd Vdd PMOS W=8 L=2 GEO=1
+ AS='(8)*6.1' AD='(8)*3.85'
+ PS='3.00*(8)+2*6.1' PD='2.00*(8)+2*3.85'
+ NRD='3.85/(8)' NRS='3.85/(8)'
MpoutQ Q X Vdd Vdd PMOS W=8 L=2 GEO=1   M=20
+ AS='(8)*6.1' AD='(8)*3.85'
+ PS='3.00*(8)+2*6.1' PD='2.00*(8)+2*3.85'
+ NRD='3.85/(8)' NRS='3.85/(8)'
MnoutQclk Q Clk net_5 Gnd NMOS W=8 L=2 GEO=1 M=4
+ AS='(8)*6.1' AD='(8)*3.85'
+ PS='3.00*(8)+2*6.1' PD='2.00*(8)+2*3.85'
+ NRD='3.85/(8)' NRS='3.85/(8)'
MnoutQx net_5 X net_6 Gnd NMOS W=8 L=2 GEO=1 M=4
+ AS='(8)*6.1' AD='(8)*3.85'
+ PS='3.00*(8)+2*6.1' PD='2.00*(8)+2*3.85'
+ NRD='3.85/(8)' NRS='3.85/(8)'
MnoutQclkbbb net_6 Clkbbb Gnd Gnd NMOS W=8 L=2 GEO=1
  M=4
+ AS='(8)*6.1' AD='(8)*3.85'
+ PS='3.00*(8)+2*6.1' PD='2.00*(8)+2*3.85'
+ NRD='3.85/(8)' NRS='3.85/(8)'
.ENDS $ hlff


.GLOBAL gnd vdd


*assumed parameters
*hdif=3.85 hdif2=6.1 cjgate=2.0 resSD=1
```

7.3.5. Example M-SAFF Deck (saff16.hsp)

```
*FILE: saff16.hsp

*SPICE netlist for "saff"
*start main CELL saff16
.SUBCKT saff16 Clk D Db Q Qb
Mrst preQ Clk preQb Vdd PMOS W=16 L=2 GEO=3
+ AS='(16)*3.85' AD='(16)*3.85'
+ PS='2.00*(16)+2*3.85' PD='2.00*(16)+2*3.85'
+ NRD='3.85/(16)' NRS='3.85/(16)'
MinM nP Db tail Gnd NMOS W=8 L=2 GEO=1
+ AS='(8)*6.1' AD='(8)*3.85'
+ PS='3.00*(8)+2*6.1' PD='2.00*(8)+2*3.85'
+ NRD='3.85/(8)' NRS='3.85/(8)'
Mlmn preQ preQb nP Gnd NMOS W=8 L=2 GEO=1
+ AS='(8)*6.1' AD='(8)*3.85'
```

```
+ PS='3.00*(8)+2*6.1' PD='2.00*(8)+2*3.85'
+ NRD='3.85/(8)' NRS='3.85/(8)'
Mtail tail Clk Gnd Gnd NMOS W=16 L=2 GEO=1
+ AS='(16)*6.1' AD='(16)*3.85'
+ PS='3.00*(16)+2*6.1' PD='2.00*(16)+2*3.85'
+ NRD='3.85/(16)' NRS='3.85/(16)'
MinP nM D tail Gnd NMOS W=8 L=2 GEO=1
+ AS='(8)*6.1' AD='(8)*3.85'
+ PS='3.00*(8)+2*6.1' PD='2.00*(8)+2*3.85'
+ NRD='3.85/(8)' NRS='3.85/(8)'
Mlpn preQb preQ nM Gnd NMOS W=8 L=2 GEO=1
+ AS='(8)*6.1' AD='(8)*3.85'
+ PS='3.00*(8)+2*6.1' PD='2.00*(8)+2*3.85'
+ NRD='3.85/(8)' NRS='3.85/(8)'
Mlpp preQb preQ Vdd Vdd PMOS W=8 L=2 GEO=1
+ AS='(8)*6.1' AD='(8)*3.85'
+ PS='3.00*(8)+2*6.1' PD='2.00*(8)+2*3.85'
+ NRD='3.85/(8)' NRS='3.85/(8)'
MrstM preQb Clk Vdd Vdd PMOS W=16 L=2 GEO=1
+ AS='(16)*6.1' AD='(16)*3.85'
+ PS='3.00*(16)+2*6.1' PD='2.00*(16)+2*3.85'
+ NRD='3.85/(16)' NRS='3.85/(16)'
MrstP preQ Clk Vdd Vdd PMOS W=16 L=2 GEO=1
+ AS='(16)*6.1' AD='(16)*3.85'
+ PS='3.00*(16)+2*6.1' PD='2.00*(16)+2*3.85'
+ NRD='3.85/(16)' NRS='3.85/(16)'
Mlmp preQ preQb Vdd Vdd PMOS W=8 L=2 GEO=1
+ AS='(8)*6.1' AD='(8)*3.85'
+ PS='3.00*(8)+2*6.1' PD='2.00*(8)+2*3.85'
+ NRD='3.85/(8)' NRS='3.85/(8)'
MnoutQ Q nQ Gnd Gnd NMOS W=8 L=2 GEO=1 M=2
+ AS='(8)*6.1' AD='(8)*3.85'
+ PS='3.00*(8)+2*6.1' PD='2.00*(8)+2*3.85'
+ NRD='3.85/(8)' NRS='3.85/(8)'
MpoutQ Q preQb Vdd Vdd PMOS W=16 L=2 GEO=1 M=2
+ AS='(16)*6.1' AD='(16)*3.85'
+ PS='3.00*(16)+2*6.1' PD='2.00*(16)+2*3.85'
+ NRD='3.85/(16)' NRS='3.85/(16)'
MpoutQb Qb preQ Vdd Vdd PMOS W=16 L=2 GEO=1 M=2
+ AS='(16)*6.1' AD='(16)*3.85'
+ PS='3.00*(16)+2*6.1' PD='2.00*(16)+2*3.85'
+ NRD='3.85/(16)' NRS='3.85/(16)'
MnoutQb Qb nQb Gnd Gnd NMOS W=8 L=2 GEO=1 M=2
+ AS='(8)*6.1' AD='(8)*3.85'
+ PS='3.00*(8)+2*6.1' PD='2.00*(8)+2*3.85'
+ NRD='3.85/(8)' NRS='3.85/(8)'
Xinv235 preQ nQ inv WP=12 LP=2 WN=4 LN=2
Xinv242 preQb nQb inv WP=12 LP=2 WN=4 LN=2
Xinv249 Q Qb inv WP=16 LP=2 WN=8 LN=2
```

```
Xinv256 Qb Q inv WP=16 LP=2 WN=8 LN=2
.ENDS $ saff16
.GLOBAL gnd vdd
*assumed parameters
*hdif=3.85 hdif2=6.1 cjgate=2.0 resSD=1
```

CHAPTER 8

STATE-OF-THE-ART CLOCKED STORAGE ELEMENTS IN CMOS TECHNOLOGY

This chapter presents clocked storage elements used in state-of-the-art micropro-cessors. MSLs, pulsed latches, and flip-flops represent the fundamental structures that are used as a baseline for derivation of circuits with extra features, such as internal clock gating, low-swing clock, or double-edge triggering. The design style and operation of each circuit implementation is discussed in detail. The chapter ends with a comparison, and general design and application recommen-dations of each circuit topology.

8.1. MASTER–SLAVE LATCH EXAMPLES

8.1.1. Derivation of Master–Slave Latch

Most commonly the MSL is built from two transmission-gate (TG) latches. There are several latch circuits that can be used in the implementation. The simplest one is the latch shown in Fig. 8.1a. The problem with this latch is that its storage node, S, appears dynamic because there is no pull-down transistor, which makes the latch susceptible to noise. A basic static version of this latch is shown in Fig. 8.1b, where a pull-down n-MOS device is added to the latch of Fig. 8.1a. The TG n-MOS transistor is a weak pull-up device, since a logic 1 has reduced swing, $V_{DD} - V_{TH}$. Also, there is a conflict between the TG n-MOS transistor and the feedback transistors during both pull-up and pull-down on the node S. These problems are remedied in the circuit shown in Fig. 8.1c. An extra TG in the feedback avoids the simultaneous pull-up/down problem, while an additional p-MOS transistor of the input TG enables good, full-swing, pull-up on node S. The latch's robustness to noise in Fig. 8.1c is therefore traded off for an increase in clocking energy.

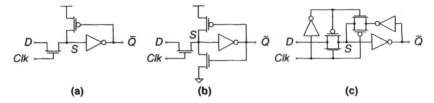

Figure 8.1. Transmission gate latches.

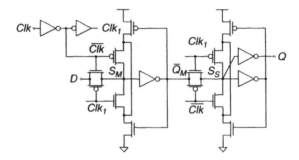

Figure 8.2. MSL with unprotected input. (Gerosa et al. 1994), Copyright © 1994 IEEE.

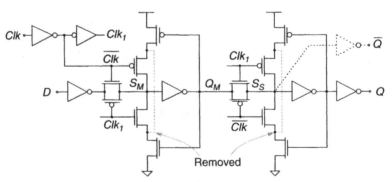

Figure 8.3. MSL with input gate isolation. (Markovic et al. 2001), Copyright © 2001 IEEE.

Usually, the conventional MSL shown in Fig. 8.2 is obtained from the latch shown in Fig. 8.1c. The energy consumption of this MSL can be reduced if the wire connecting the drains of the top p-MOS and bottom n-MOS transistors is removed, as shown in Fig. 8.3.

Circuit Operation When the clock (*Clk*) is low, the TG of the master latch is transparent and input data *D* are stored on the master's latch storage node, S_M. The output $\overline{Q}_M$ of the master latch follows S_M and stores its inverse. The feedback of the master latch is turned off, while the feedback of the slave latch is turned on, holding the previously stored value at the slave's storage node, S_S. When *Clk* goes from low to high, the TG of the master latch becomes opaque, the master latch's feedback closes up, keeping the stored value of $\overline{Q}_M$. The slave latch is transparent, and the output of the master latch, $\overline{Q}_M$, is passed to the slave latch and stored on its storage node, S_S. This newly stored value of S_S is inverted and passed to the output Q of the latch.

Noise Robustness   The master latch of the circuit in Fig. 8.2 is susceptible to the input charge injection. Noise sources that affect the latch state node are illustrated in Fig. 8.4.

The latch in Fig. 8.2 is dominantly sensitive to the first noise source. If the wire driving the line is long, the neighboring line can capacitively couple to the latch input wire and introduce a negative spike (below ground) that will turn on the master TG that is nominally off, and the value stored in the master latch will be lost. This can be overcome by the input gate isolation as shown in Fig. 8.3. In the figure it is shown as an inverter, although it can be any logic gate that is close to the latch input. The noise sources arising from unrelated signal coupling (cross talk) and power-supply noise are attenuated by the latch feedback that makes S_M pseudostatic. There is an additional inverter at the output of the circuit in Fig. 8.3 for noninverting operation (shaded inverter). The complementary output can be easily generated by the addition of one extra inverter, as shown in Fig. 8.3 (dashed inverter).

Starting from the MSL in Fig. 8.2, a number of improvements can be made, resulting in the structure shown in Fig. 8.3. Removal of the wire allows for a more efficient layout due to the reduction of contact holes when the TGs are replaced with series-connected switches (Suzuki et al. 1973). The slave latch has the same structure as the master latch with the addition of an extra inverter that drives the output and prevents loading of the feedback loop by the output capacitance. A similar circuit was proposed by Gerosa et al. (1994).

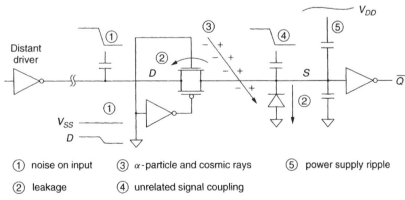

① noise on input   ③ α-particle and cosmic rays   ⑤ power supply ripple

② leakage   ④ unrelated signal coupling

Figure 8.4. Sources of noise affecting the latch state node. (Partovi in Chandrakasan et al. 2001), Copyright © 2001 IEEE.

8.1.2. C²MOS Master–Slave Latch

The C²MOS MSL, (Suzuki et al. 1973) is based on the C²MOS latch shown in Fig. 8.5b. The C²MOS latch is obtained from the TG latch in Fig. 8.5a when the wire that connects the drains of the top p-MOS and bottom n-MOS transistors is removed from the TG latch. The addition of a weak gated feedback enables the pseudostatic operation of the C²MOS MSL, as shown in Fig. 8.6.

The critical D–Q path of the C²MOS MSL is shortened by placing the feedback loop outside the path from D to Q. This makes this latch faster than a conventional MSL with input gate isolation.

8.1.3. Comparison

Figure 8.7 shows the comparison of the timing and energy parameters in the MSL and C²MOS latches. In this particular example, the C²MOS latch had larger clocked transistors, resulting in energy that is twice as large, as illustrated

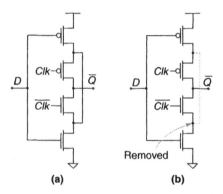

Figure 8.5. Dynamic latches with gate isolation: (a) transmission gate, (b) C²MOS.

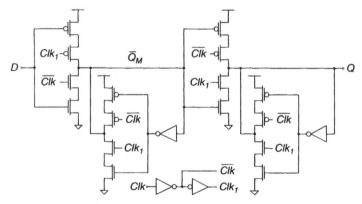

Figure 8.6. C²MOS latch (C²MOS). (Suzuki et al. 1973), Copyright © 1973 IEEE.

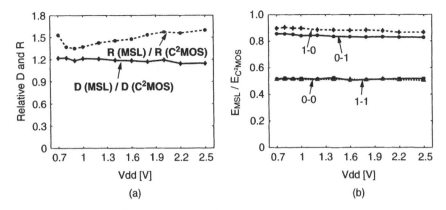

Figure 8.7. Comparison of MSLs: (a) timing, (b) energy.

in Fig. 8.7b (transitions 0–0 and 1–1). The larger clocked transistors resulted in reduced Clk-Q delay, which led to both shorter delay and shorter internal race immunity of the C^2MOS latch. Figure 8.7a shows that the MSL has about 40% better internal race immunity and about 20% worse delay for this particular sizing. Both of these circuits have moderate delay, which is a general property of MSLs, so they are suitable for noncritical paths. In this particular study, MSL would be favorable due to larger internal race immunity and smaller energy consumption. Both latches were loaded with a small load, corresponding to approximately eight minimum-sized inverters. The comparison results can change if the size of the clocked transistors is fixed or the latches optimized for a different output load.

8.2. FLIP-FLOP EXAMPLES

8.2.1. Hybrid-Latch Flip-Flop

The HLFF by Partovi et al. (1996) is shown in Fig. 8.8. It is a single-input single-output, positive edge-triggered flip-flop. Its derivation is explained in depth in Chapter 2. This design initiated a whole series of similar devices, representing the limited comeback of flip-flop-based clocking, because the CSEs required very low overhead and because of the increasing importance of clock uncertainty absorption.

Circuit Operation   Prior to the rising edge of *Clk*, the circuit is in the *precharge phase*, node $\overline{S}$ is precharged to high, and the output inverters hold the previously stored logical value on $\overline{Q}$, which is decoupled from $\overline{S}$ because the second stage is off.

At the rising edge of *Clk*, the pull-down side of both the first and the second stage is enabled for a period of time defined by the three-inverter delay chain. During this period the flip-flop is transparent and *D* can be sampled into the

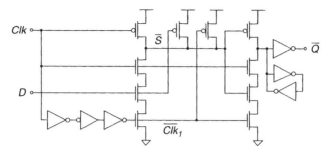

Figure 8.8. Hybrid-latch flip-flop. (Partovi et al. 1996), Copyright © 1996 IEEE.

flip-flop. If D is high in this period, node $\overline{S}$ goes low, forcing $\overline{Q}$ low. If D is low, $\overline{S}$ goes high and $\overline{Q}$ goes to high. Once $\overline{Clk}_1$ goes low, node $\overline{S}$ is decoupled from D and is held at high or the P-MOS device that is driven by $\overline{Clk}_1$ begins to precharge it to high. The falling edge of $\overline{Clk}_1$ is the latching edge for the pull-down path of the second stage, while $\overline{S}$ rising is the latching edge for the pull-up path of the second stage. At the falling edge of Clk, node $\overline{S}$ is precharged to high by the p-MOS transistor that is driven by Clk, and $\overline{S}$ remains high as long as the Clk stays low.

8.2.2. Semidynamic Flip-Flop

The semidynamic flip-flop (SDFF) by Klass (1998), is shown in Fig. 8.9. It is a single-input single-output, positive edge-triggered flip-flop. The domino-style front end allows for efficient embedded combinational logic and reduces the load on the data network.

Circuit Operation The SDFF is composed of a dynamic front end and a static back end. When Clk is low, the circuit is in the *precharge phase*. Node $\overline{S}$ is precharged high and node Q is decoupled from the first stage. The output inverters hold the previous values of Q and $\overline{Q}$. The *evaluation phase* begins at the rising edge of Clk. If D is low, $\overline{S}$ remains high, driving Q low and $\overline{Q}$ high.

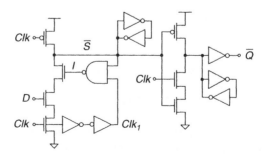

Figure 8.9. Semidynamic flip-flop. (Klass 1998), Copyright © 1998 IEEE.

With D high, $\bar{S}$ will discharge, driving Q high and $\bar{Q}$ low. Three gate delays after the rising edge of *Clk*, the output, I, of the NAND gate goes low, preventing discharge of node $\bar{S}$ by subsequent late 0–1 transitions on D. The narrow capture pulse makes this circuit appear edge-triggered. It is worth noting that a glitch occurs at Q when $D = Q =$ high, as shown in Fig. 8.10. This problem also occurs in the HLFF design. If proper caution is not exercised during transistor sizing, this glitch can cause the output latch to change the state. The glitch also increases power consumption.

A more systematic approach in derivation of the structure through Karnough's maps eliminates this problem, as shown by the full realization of SDFF in Chapter 2.

8.2.3. Sense-Amplifier-Based Flip-Flop

The SAFF (Matsui et al. 1994; Montanaro et al. 1996) is shown in Fig. 8.11. It is a differential-input differential-output, positive edge-triggered flip-flop. It consists of a pulse-generating stage implemented as a precharged sense-amplifier and $S-R$ latch implemented with two cross-coupled NAND gates. Although the pulse-generating stage was discussed in detail in Chapter 2, we will address some of the issues of this particular implementation in this section. Then we will focus on this flip-flop's speed bottleneck, and show the series of proposed methods to improve the speed of this device.

Circuit Operation   This flip-flop operates in the precharge/evaluate mode. When *Clk* is low, the flip-flop is in the *precharge phase*, the input differential-pair is off, and the cross-coupled NAND gates ($S-R$ latch) in the output stage hold the previously stored logic value at Q and $\bar{Q}$.

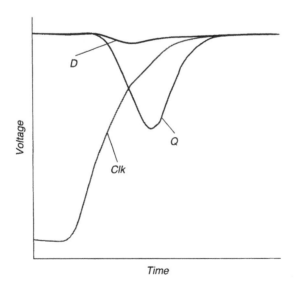

Figure 8.10. Illustration of SDFF output glitch.

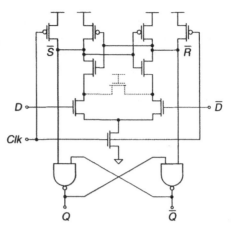

Figure 8.11. Sense-amplifier-based flip-flop. (Montanaro et al. 1996), Copyright © 1996 IEEE.

At the rising edge of Clk, the differential pair senses complementary inputs, D and $\overline{D}$, and generates a low pulse on either $\overline{S}$ or $\overline{R}$. The $S-R$ latch captures the logic values of the differential inputs and holds them until the next rising edge of the *Clk*. During the evaluation phase, either the $\overline{S}$ or $\overline{R}$ is low, discharged by the pull-down path conditioned by the *Clk* and D or $\overline{D}$. If the data value changes after the pulse was generated, the pull-down path can be turned off, leaving the discharged node $\overline{S}$ (or $\overline{R}$) floating. In order to prevent the floating node from being charged by the leakage or coupling noise, the alternate pull-down path is provided with the addition of the weak n-MOS pass gate, shown in dotted lines in Fig. 8.11. The primary role of this pass-gate is to staticize nodes $\overline{S}$ or $\overline{R}$ when there is a glitch on the data inputs that follows the rising edge of the clock (Montanaro et al. 1996). It also helps equalize the voltage values of the two differential branches during precharge and minimizes the effect of the previously evaluated data values on the following evaluation. The alternate methods of providing static operation and enhancements to the pulse-generator stage, including its formal derivation, are covered in detail in Chapter 2.

Evolution of the S–R Latch in the Second Stage   The pulse-generating stage exhibits a very small delay and setup time, due to its sense-amplifier implementation, which incorporates positive feedback. Two cross-coupled NAND gates in the $S-R$ latch present the speed bottleneck of this flip-flop. In the worst case, the signal has to propagate through both NAND gates until it reaches the output of the flip-flop. More precisely, the falling transition on any of the outputs is slower than the rising transition by one NAND gate delay. As the overhead of the CSE became more and more important in the devices used on critical paths, the SAFF was seen as a structure with great potential for low overhead, due to the fast input-sensing stage. The NAND-based $S-R$ latch was identified as the main bottleneck and became the focus of the efforts to improve the speed of this

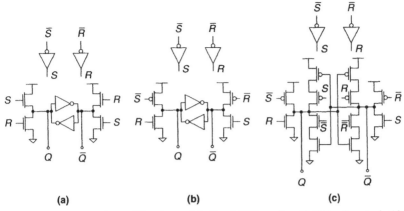

Figure 8.12. $S-R$ Latch modifications: (a) all-n-MOS push-pull (Gieseke et al. 1991); (b) complementary push-pull (Oklobdzija and Stojanovic 2001); (c) complementary push-pull with gated keeper (Nikolic et al. 1999).

flip-flop. In Fig. 8.12 we have summarized the $S-R$ latch modifications proposed over the years.

The first modification to the classic cross-coupled NAND stage, shown in Fig. 8.12a, was accomplished using symmetric push-pull logic (Gieseke et al. 1991), to implement a symmetric $S-R$ latch stage. A flip-flop with this output stage was used in the critical paths of an Alpha 21264 processor (Partovi in Chandrakasan et al. 2001). One of the major drawbacks of this all-n-MOS push-pull scheme are the n-MOS source-followers that have a weak pull-up capability, relying on the keeper (cross-coupled inverters at the output) to finish charging the output node once the state is changed. The second modification, using a complementary push-pull scheme (Oklobdzija and Stojanovic 2001), achieves significant speed improvement over the first one by decoupling the role of the push-pull driver from that of the keeper. In the circuit in Fig. 8.12b, the $\bar{S}$ and $\bar{R}$ signals directly drive the p-MOS drivers, while S and R are generated from skewed inverters (S and R rising much faster than falling). Transistors should be sized such that the delay of the p-MOS driver is equal to that of the skewed inverter plus the n-MOS driver. In this arrangement drivers are capable of fully switching the output signal from rail to rail, without needing help from the keeper. The third modification (Nikolic et al. 1999), shown in Fig. 8.12c, introduced the gated keepers in order to prevent the conflict between the driver and the keeper during the switching of the output. This can potentially speed up the $S-R$ latch, but slows down the first stage because of the additional control ports on the keeper that load the first stage directly. Depending on the size of the keeper device, the $S-R$ latch versions with and without the gated keeper result in better performance.

8.2.4. Modified Sense-Amplifier-Based Flip-Flop

The mismatch between rising and falling $Clk-Q$ delays of the SAFF is solved using a symmetric $S-R$ latch, as discussed in the previous section. The M-SAFF,

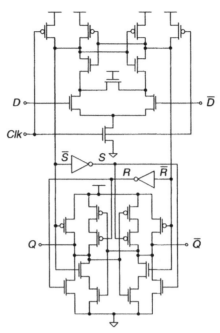

Figure 8.13. Modified sense-amplifier-based flip-flop. (Nikolic et al. 1999), Copyright © 1999 IEEE.

shown in Fig. 8.13, is composed of the sense-amplifier pulse-generating stage and the symmetric $S-R$ latch with a gated keeper.

Circuit Operation   The overall circuit operation is identical to the operation of the SAFF. Inputs $\overline{S}$ and $\overline{R}$ are the set and reset inputs of the $S-R$ latch, respectively. The low level at both of these inputs is not allowed, which is ensured by the sense amplifier. In this implementation of the $S-R$ latch circuit, both Q and $\overline{Q}$ change simultaneously, unlike in the cross-coupled NAND version. The inverter-like drivers directly drive the output load, while the output keeper inverters are gated. This arrangement prevents transient energy dissipation between the driver and the keeper inverters. During the precharge phase, the driver is disabled by the sense-amplifier signals $\overline{S}$ and $\overline{R}$ and the keeper retains the state at the output of the flip-flop. There exists a the trade-off in keeper sizing, between the delay of the flip-flop affected by the parasitic load of the keeper and the length of the wire that the structure can drive due to the requirement to absorb coupling noise without a change in state.

8.2.5. Comparison

Now that the principles of operation have been explained, the representative structures can be compared in terms of delay and energy. The minimum $D-Q$ performance is illustrated in Fig. 8.14 for some average CSE load, as explained in

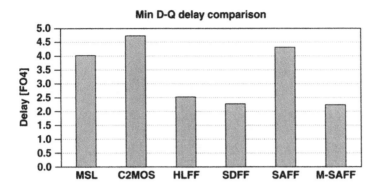

Figure 8.14. CSE delay comparison (0.18 μm, high load).

Chapter 7. Delay is expressed in terms of the FO4 inverter delay, which is shown to be relatively independent of technology scaling. The typical clock-cycle overhead due to CSE is 2–4 FO4s. The performance advantage of flip-flops (HLFF, SDFF, SAFF, M-SAFF) over MSLs (MSL, C^{2}MOS) is due to the negative setup time of the structures with a transparency window when compared to the positive setup time of M-S elements. Due to internal pulse generation, flip-flops can have a narrower capturing pulse than externally Pulsed Latches (PLs). These latches cannot have an arbitrarily wide clock pulse (hence large negative setup time), due to the hold-time restrictions, so in delay performance they are close to well-designed flip-flops. It is worth noting that these are just some of the state-of-the-art designs that are chosen to illustrate some of the key points in high-performance CSE design, and not necessarily to represent the fastest/lowest energy structures available.

In modern processor design, the energy of the CSEs is a very important parameter. Following the energy breakdown definitions in Chapter 3.2, Fig. 8.15 illustrates different components of energy dissipation, which illustrate advantages/weaknesses in the design of the representative CSEs. There are several very important conclusions that we can draw from the illustrated energy components. For one, the MSLs are inherently two-phase elements, and the energy needed to generate the second clock phase is either attributed to the external clock energy (the energy parameter illustrating the load on the clock distribution network) or to the circuitry inside the CSE (internal clock energy). In the case of SDFF and HLFF, energy is dissipated in every clock cycle in the pulse generator circuitry. Differential structures using precharge dissipate the energy charged in the majority of nodes in every clock cycle, such as SAFF or M-SAFF.

In low-power designs, the performance of a CSE is properly assessed only if the structure is evaluated versus supply-voltage scaling, since that is one of the most significant ways to save energy. It is important that CSE performs well and has robust behavior for a range of supply voltages. Figure 8.16 presents a comparison of minimum $D-Q$ delay and internal race immunity, R, of flip-flops. It is interesting that the relative flip-flop circuit delay analyzed in this example

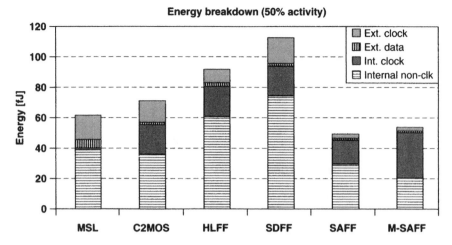

Figure 8.15. CSE energy breakdown (0.18 μm, high load).

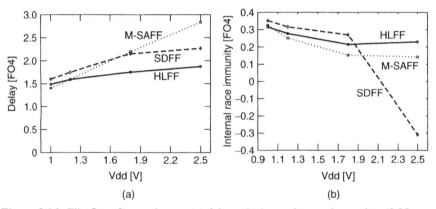

Figure 8.16. Flip-flop Comparisons: (a) delay, (b) internal race immunity (0.25 μm, light load).

reduces the supply voltage with scaling. This is because of the favorable scaling of stack transistors with reduced body effect at lower supplies, as in HLFF and SDFF, or because of the positive-feedback cross-coupled differential circuits in M-SAFF. All of these circuits show a very small race margin, but this is usually not a concern, since the fast flip-flop circuits are placed on the critical paths. Aggressive clock-skew specification requires careful clock distribution and deskewing circuits leading to the increased energy consumed in the clock distribution network.

In low-energy systems, energy is the primary concern. Figure 8.17 shows a comparison of energy-per-transition in representative MSLs and flip-flops optimized for a light output load. All the results are under scaled supply voltage,

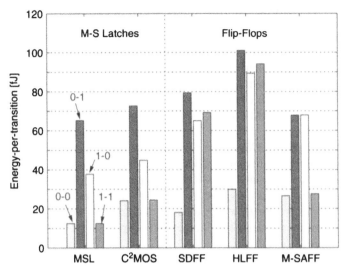

Figure 8.17. CSE energy-per-transition (0.25 μm, light load). (Markovic et al. 2001), Copyright © 2001 IEEE.

$V_{DD} = 1$V in 0.25 μm technology. The results are different from the case when the same set of circuits is optimized for high output load and at a nominal supply voltage provided by technology. Most notably, SAFF and M-SAFF optimized for light load exhibit higher energy consumption than any of the MSLs, contrary to the high-load case shown in Fig. 8.15. MSLs exhibit the lowest energy consumption among the circuits analyzed in this particular example. The MSLs are therefore the preferable CSEs in low-energy applications. If the performance is not the most important goal, MSLs can be used on the critical paths as, for example, in the PowerPC 603 (Gerosa et al. 1994).

High-performance applications need greater speed and are forced to accept the fast CSE structures with high-energy requirements. In low-energy applications or in noncritical paths, MSLs are preferred over flip-flops, because the MSLs have better internal race immunity and require lower energy consumption, at the expense of a small increase in delay. In the following sections, techniques are presented that further reduce the energy dissipated in CSEs.

8.3. CLOCKED STORAGE ELEMENTS WITH LOCAL CLOCK GATING

This technique attempts to minimize the amount of internal clock energy dissipated in the CSE, which was shown in Fig. 8.15 to be one of the larger components in energy breakdown. The gating mechanism turns the internal clock off when input and output data are equal. Since there is a cost associated with local clock gating, the use of these CSEs is justified for low switching activities of the input data. The internal clock-gating technique can be applied to all circuits introduced thus far.

8.3.1. Master–Slave Latch with Local Clock Gating

The internal clock gating is applied to MSL, with the idea of further reducing its energy consumption and maintaining its good internal race immunity. The gated MSL (G-MSL) with internal clock gating is shown in Fig. 8.18 (Markovic et al. 2001). This latch is derived from MSL. The circuitry for internal clock gating, which is similar to the clock gating circuitry presented by Strollo et al. (2000), is surrounded by dashed lines in Fig. 8.18.

Circuit Operation Comparator (comp) performs an XNOR operation on D and Q. The comparator is implemented in the complementary pass-transistor logic (CPL) technique by Yano et al. (1990), taking advantage of the freely available true and complementary signals. This reduces the transistor count of the clock gating circuitry. When $D \neq Q$, output of the comp is low and enables external *Clk* to propagate through the internal clock generation circuits that generate internal clocks $\overline{Clk}$ and Clk_1.

The pull-up side of the input clock inverter is chosen to be gated because the CPL realization of an XNOR has better pull-down, allowing for faster generation of the internal clocks than if the pull-down side of the input inverter was gated. Weak feedback is added around the inverter that outputs Clk_1 for the pseudostatic operation.

Compared to the conventional MSL (CMSL), the circuit of Fig. 8.18 achieves lower energy consumption when the switching activity of the data input is less than 0.3, as shown in Fig. 8.19b. The logic for internal clock gating incurs delay cost, which is reflected in the increased setup time of the gated latch, as shown in Fig. 8.19a. Internal race immunity in general is not affected by the gating operation because of similar variations in the Clk-Q delay and hold time.

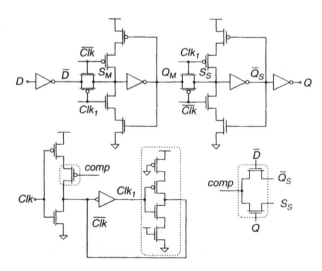

Figure 8.18. Gated MSL. (Markovic et al. 2001), Copyright © 2001 IEEE.

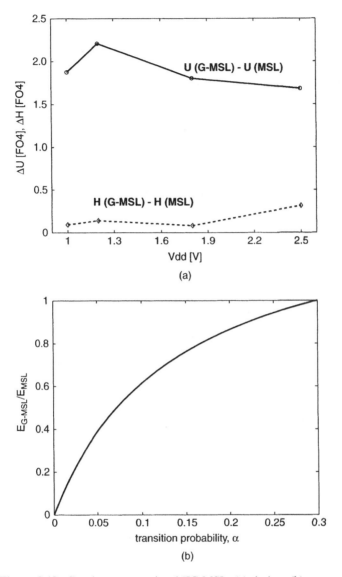

Figure 8.19. Gated vs. conventional TG MSL: (a) timing, (b) energy.

8.3.2. Data-Transition Look-Ahead Latch

The DTLA-L by Nogawa and Ohtomo (1998) is shown in Fig. 8.20. It is a noninverting PL.

Circuit Operation   The data-transition look-ahead (DTLA) circuit performs an XNOR function on D and Q. When $D = Q$, the DL circuit produces a logic 0

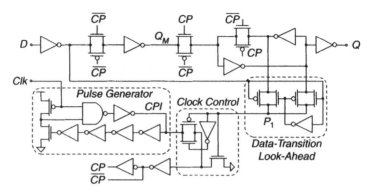

Figure 8.20. Data-transition look-ahead latch. (Nogawa and Ohtomo 1998), Copyright © 1998 IEEE.

at P_1 and generation of the internal clock $\{CP, \overline{CP}\}$ is disabled. When $D \neq Q$, P_1 is low and the CC circuit enables generation of $\{CP, \overline{CP}\}$.

The PG circuit generates a short pulse, CPI, at every rising edge of the external clock, Clk. Internal clock pulse CP then triggers the latch if $D \neq Q$. The PG is essential for the operation of the latch. If there were no pulse generator, this latch could be triggered by data instead of the clock. For example, if $D \neq Q$ and the rising edge of CPI arrives, then the clock pulse CP is generated and Q changes. However, if D changes again while the clock is still high and becomes different from Q, then another internal clock, pulse CP, would be generated and the CSE would be actually triggered by the data.

Analysis of DTLA-L   In order to evaluate the benefit of clock gating, it is essential to find the energy cost associated with the internal clock gating circuitry. For that purpose, a portion of the DTLA-L circuit shown in Fig. 8.21 is analyzed.

When the energy-per-transition of the circuit in Fig. 8.20 is subtracted from the energy-per-transition of the circuit in Fig. 8.21, the energy cost in data look-ahead, clock control, and pulse generator is obtained. This only applies to 0–1 and 1–0 input transitions, because only then are all subcircuits in Fig. 8.20 and Fig. 8.21 active. For 0–0 and 1–1 input transitions, the internal clock (shaded inverters) is activated in the circuit of Fig. 8.21 and deactivated in the circuit in Fig. 8.20.

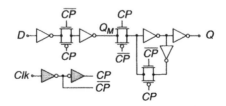

Figure 8.21. DTLA-L without gating.

Clk

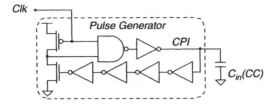

Figure 8.22. Pulse generator.

The PG is commonly shared among several latches, so further energy break-down is needed to understand exactly where the energy goes. The energy consumed by the PG is estimated by simulation of a stand-alone PG loaded with capacitance $C_{in}(CC)$ that *CPI* sees when looking into the CC circuit, as shown in Fig. 8.22. The PG consumes energy regardless of what input transition occurs. A portion of the PG's energy dissipation is attributed to each CSE through external clock energy parameter.

Energy Efficiency of DTLA-L The energy saving capabilities of DTLA-L depend on two parameters: (1) number of latches, N, driven by a single PG, and (2) input data-transition probability. The energy consumed per latch during one clock cycle, when $D = Q$, is given by:

$$E_{D\text{-}idle} = E_{0-0} = E_{1-1} = \frac{E_{PG}(N)}{N} + E_{C_{in}} \tag{8.1}$$

The energy consumption when D undergoes a 0–1 or 1–0 transition is given by Eqs. (8.2)–(8.3):

$$E_{0-1} = E_{D\text{-}idle} + E_{Clk} + E_{DL+CC} + E_{int} + E_{ext} \tag{8.2}$$

$$E_{1-0} = E_{D\text{-}idle} + E_{Clk} + E_G + E_{int} \tag{8.3}$$

Comparison with M-S Latch Assuming there is no glitching at input D, the probability of 0–1 and 1–0 transitions is equal: $\alpha_{0-1} = \alpha_{1-0} = \alpha/2$, where α is data-transition probability. Under this postulation, the average energy consumption of the DTLA-L and the conventional MSL CMSL are:

$$E_{DTLA-L} = \frac{\alpha}{2} \cdot (E_{0-1} + E_{1-0}) + \frac{1-\alpha}{2} \cdot E_{D-idle} \tag{8.4}$$

$$E_{MSL} = \frac{\alpha}{2} \cdot (E_{0-1} + E_{1-0}) + \frac{1-\alpha}{2} \cdot E_{Clk} + E_{C_{in}} \tag{8.5}$$

In Eq. (8.5), $E_{C_{in}}$ is the energy consumed in switching the $C_{in(Clk)}$ of the MSL. This term represents the energy consumed by the simple clock buffer that drives

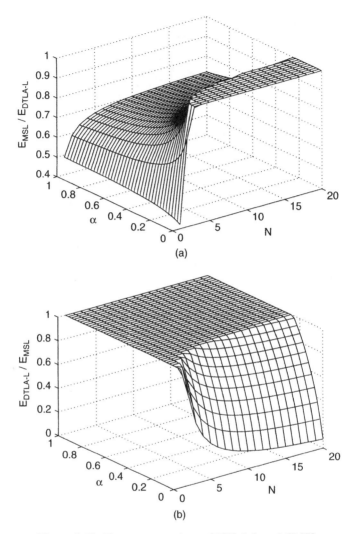

Figure 8.23. Energy comparison of DTLA-L and CMSL.

$C_{in(Clk)}$ and is included for a fair comparison with the DTLA-L where $E_{PG}(N)/N$ represents the energy consumption in PG per latch.

Figure 8.23 shows comparison of energy consumption in DTLA-L and MSL as a function of N and α. The figure shows that DTLA-L has better energy efficiency than the MSL for N > 2 and $\alpha < 0.25$.

8.3.3. Clock-on-Demand Pulsed Latch

The clock-on-demand PL (COD-PL) by Hamada et al. (1999) is shown in Fig. 8.24. It is a positive edge-triggered, noninverting PL. The circuits enclosed

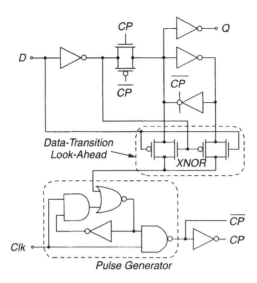

Figure 8.24. Clock-on-demand PL. (Hamada et al. 1999), Copyright © 1999 IEEE.

with dashed lines show the cost associated with pulse generation and data-transition look-ahead.

Circuit Operation   As in the DTLA-L, the data-transition look-ahead circuit also performs an XNOR function on D and Q. When $D = Q$, XNOR $= 0$ and CP is disabled. When $D \neq Q$, the PG circuit generates a short pulse, CP, at every rising edge of the external clock, *Clk*. The pulse ends only when XNOR changes to high, which means that its duration is proportional to the delay of the transmission gate, inverter, XNOR, and the PG logic.

Unlike the DTLA-L, the COD-PL has its local pulse generation. As pointed out by Hamada et al. (1999), this helps avoid problems with distortion of the pulse in the clock distribution and the power penalty of the pulse clock generator. The clock control function is integrated in the internal pulse generator of the COD-PL, as opposed to the DTLA-L. This reduces the area cost of the COD-PL and promises better energy efficiency than in the DTLA-L.

Energy-Efficient Pulse-Generator   Careful optimization of PG is the key to the minimization of the energy overhead associated with internal clock gating. If the PG circuit were implemented in complementary CMOS, there would be energy consumption in the PG even when D is idle, as illustrated in Fig. 8.25. In order to avoid this unnecessary energy consumption, AND and NOR circuits are implemented as a single compound gate, as shown in Fig. 8.26.

Comparison with MSL   The framework used in the analysis of the energy efficiency presented in the DTLA-L example is also applied to COD-PL. Extending our analysis here a step further and exploring the impact of circuit sizing on

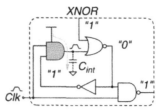

Figure 8.25. Straightforward implementation with CMOS gates (energy inefficient).

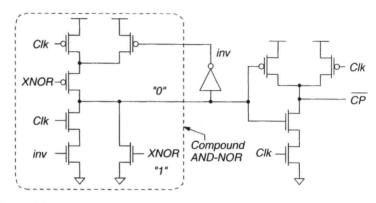

Figure 8.26. Energy-efficient implementation with compound AND–NOR gate.

the energy efficiency, we conclude that the internal clock gating technique is rarely effective in low-energy designs. This is illustrated in the example of the COD-PL where the transistor sizes are optimized for two cases: high-speed and low-energy. The sizes of the MSL are optimized accordingly as well. Figure 8.27 shows energy consumption in COD-PL relative to the energy consumed by the MSL versus transition probability α, where $\alpha_{0-1} = \alpha_{1-0} = \alpha/2$. Since in high-performance circuits clock transistors are large (4× sizing), this technique promises to save energy, as shown in Fig. 8.27. This is because the clock gating logic represents a small portion of the overall circuit area. However, in low-energy CSEs (1× sizing) with small clock transistors this technique is not as effective as depicted in Fig. 8.27.

8.3.4. Conditional Capture Flip-Flop

The CCFF by Kong et al. (2000) is shown in Fig. 8.28. It is a positive edge-triggered differential-input differential-output flip-flop. As discussed in Chapter 6, this circuit is in essence a $J-K$ flip-flop.

Circuit Operation   The CCFF uses the capturing latch of the M-SAFF (Oklob-dzija and Stojanovic 2001), in addition to the internal clock gating in the PG stage.

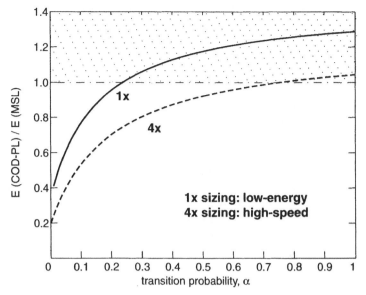

Figure 8.27. Impact of circuit sizing on the energy efficiency of COD-PL. (Markovic et al. 2001), Copyright © 2001 IEEE.

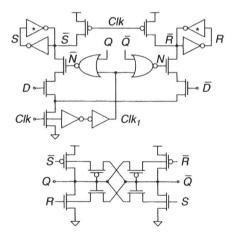

Figure 8.28. Conditional capture flip-flop. (Kong et al. 2000), Copyright © 2000 IEEE.

When *Clk* is low, the flip-flop is in the *precharge* phase, $\overline{S}$ and $\overline{R}$ are precharged high, and the S–R latch is disabled. At the rising edge of *Clk*, the behavior of the CCFF depends on the incoming data value — if the new data are not equal to the previously recorded output data, one of the outputs of the NOR gates is high, enabling pull-down of $\overline{S}$ or $\overline{R}$. The transparency period of the differential pair is the sum of two inverter delays and one NOR gate delay long because N and $\overline{N}$

both go low when Clk_1 is high. During this short transparency period, new data are latched by the $S-R$ latch at the output.

8.3.5. Comparison

The delay of the gated MSL G-MSL is increased relative to the conventional latch, as discussed before, due to an increase in setup time. As in the G-MSL, in the DTLA-L and COD-PL, setup and hold times are affected by the delay of internal gating logic and also by the width of the internal clock pulse, resulting in the delay increase in these CSEs. For example, G-MSL has about a 2FO4 larger delay than standard MSL, as shown in Fig. 8.29a. It should be noted, though, that internal clock gating does not always result in delay degradation. This is in cases where the internal clock-gating logic is outside the Clk-Q path, as in CCFF. The simulation results for this circuit were not available to us at the time of writing. Figure 8.29b contains a comparison of the internal race immunity of the gated and the conventional MSL. The G-MSL has an even better internal race margin than the conventional MSL because of its increased Clk-Q delay. Figure 8.29b also confirms the general trend: flip-flops and PLs have the smallest internal race immunity, R, MSLs have large R, and MSLs with internal clock gating have the largest R.

The average energy consumption of the CSEs with internal clock gating as a function of input data activity is shown in Fig. 8.30a. Since it is not fair to compare flip-flops and MSLs in terms of energy efficiency because the flip-flops have a higher performance, a comparison of the energy-delay product is given in Fig. 8.30b. From Fig. 8.30b it appears that the DTLA-L is the best latch for $\alpha \in [0.03, 0.23]$. The MSL offers better energy-delay trade-off than G-MSL for $\alpha > 0.12$.

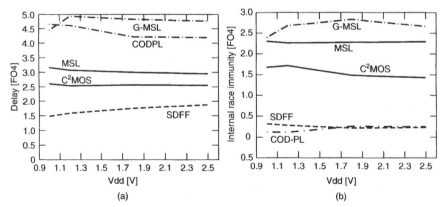

Figure 8.29. Timing parameters in latches and flip-flops with local clock gating. (Markovic et al. 2001), Copyright © 2001 IEEE.

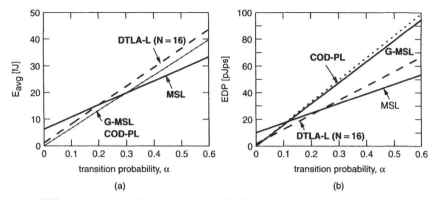

Figure 8.30. Energy and EDP in latches and flip-flops with local clock gating. (Markovic et al. 2001), Copyright © 2001 IEEE.

The alternative approach to internal clock gating is to minimize the external clock energy component. A low-swing clock is distributed to specially designed CSEs. This is the topic of the next section.

8.4. LOW-SWING CLOCK STORAGE ELEMENTS

A reduced-swing clock CSE technique targets the energy savings in the clocking of a CSE. The most effective way to employ reduced-swing clocks is to use CSEs that can operate with reduced-swing clock input and that do not require any redesign of the clock driver. This class of CSEs has n-only clocked transistors, with the clock network simply operating under the reduced supply voltage.

Standard clocked storage elements cannot be used with a low-swing clock, since any clocked p-MOS transistor will not fully turn off, causing static current and reduced robustness. It is therefore imperative to design storage elements amenable to low-swing clocking, in order to identify the topology that enables maximum energy reduction while incurring minimum delay penalty and degradation of robustness to clock noise. Reduced-swing clocking allows energy reduction at the expense of some cycle-time increase. As the clock voltage is reduced, the consumption of energy also becomes smaller, but with diminishing returns as the clocking power becomes much less than the data switching power.

8.4.1. CSE Examples

The operation of these latches is much like the conventional topologies they are derived from by eliminating p-clocked transistors (e.g., the latch in Fig. 8.31a) or adding additional transistors to improve pull-up of the state node (e.g., latch in Fig. 8.31d).

For noncritical paths, the N-only static MSL (N-MSL, Fig. 8.31a) is obtained from the standard MSL (Tschanz et al. 2001) by removing the clocked p-MOS transistors and allowing gating of only the pull-down keepers.

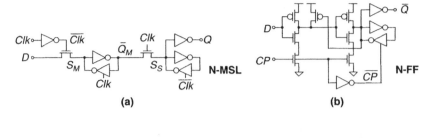

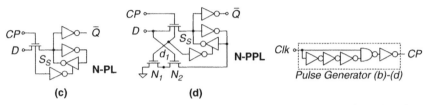

Figure 8.31. N-only clocked latches: (a) conventional TG MSL, (b) pulsed-latch, (c) conventional PL, (d) push-pull PL.

For the performance-critical paths, explicitly pulsed latches are used that allow sharing of the pulse generator. N-FF (Fig. 8.31b) is a simple flip-flop (Tschanz et al. 2001). N-PL (Fig. 8.31c) is derived from the transmission gate-based PL presented in Tschanz et al. (2001). An n-only push-pull clocked cycle latch (N-PPL, Fig. 8.31d) is constructed from N-PL by adding N_1 and N_2 for faster pull-up operation.

8.4.2. Comparison

The framework presented by Tschanz et al. (2001) is used for latch and flip-flop optimization in a 130-nm technology, with a 50-fF load at the output for all low-swing clock CSEs. A global optimizer is used to determine the sizes of all transistors that minimize the energy consumption of the CSE (for data activity of 0.1) for different delay targets. CSE delay is the sum of the worst-case Clk-Q delay and worst-case setup time, considering both logic polarities in the critical paths. The maximum input capacitance of the clock and data drivers is limited to 12.5 fF. It is assumed that the input driver is located adjacent to each CSE for robust operation.

Energy-delay comparisons of CSE designs at single high-V_{DD} and with low-swing clock (Fig. 8.32) show that while PL offers the best performance at high-V_{DD} (Tschanz et al. 2001), PFF is the preferred design for low-swing clocking. Assuming that the target delay for performance-critical CSEs is equal to 1.5 × FO3 (1.2 FO4) inverter delays, the low-swing N-PL can achieve 1.8 × FO3 (1.4 FO4) inverter delays – 20% CSE delay degradation. For noncritical CSEs, MSL is the most robust and energy-efficient at high-V_{DD} (Tschanz et al. 2001), while N-PPL is best for low-swing clocking.

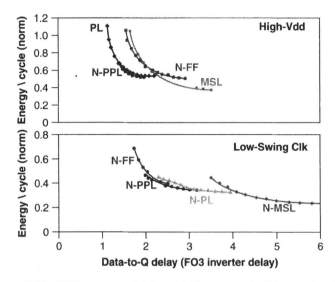

Figure 8.32. CSE energy and delay: (a) high-V_{dd} and (b) low-swing *Clk*.

Another important consideration in a design with a low-swing clock is the impact of clock noise on CSE delay (Fig. 8.33). We quantify it by measuring the increase in *Clk-Q* delay when the clock swing is reduced by the amount of noise. Among the high-performance CSEs considered here, N-FF provides the highest robustness against clock noise. All latches fail as clock noise approaches 12% of the clock voltage. Nevertheless, N-FF offers the best clock noise rejection.

N-FF[4] and N-PPL circuits are the most energy-efficient choices for performance-critical and noncritical parts of a microprocessor with a low-swing clock. They also offer better robustness against clock noise than N-CL. Full-chip energy savings of low-swing clocking are greater than those from simple V_{DD}

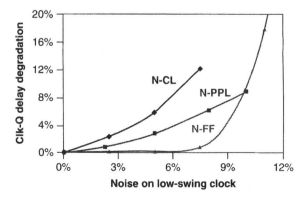

Figure 8.33. Effect of clock noise on low-swing clock latch delay.

lowering only when the power of the clocking subsystem is more than 30% of the total power.

8.5. DUAL-EDGE-TRIGGERED CLOCKED STORAGE ELEMENTS

Dual-edge-triggered (DET) CSEs sample their inputs and update their outputs on both the rising and falling edges of the clock. With this approach, the maximum toggle frequency of the clock is identical to the maximum toggle frequency of the data. In contrast, conventional single-edge-triggered clocked storage elements require twice as high clock frequency for the same data throughput. Thus, a migration from single- to dual-edge triggered clocking strategy to a first approximation halves the clock energy.

8.5.1. DET Latch-mux

DET Latch-mux (LM) by Llopis and Sachdev (1996), shown in Fig. 8.34, is the dual-edge counterpart of single-edge MSL by Gerosa et al. (1994). The basic building blocks, latches and the multiplexer, can be easily identified on the schematic. The latches are implemented using pass gates and are staticized by clocked feedback. The multiplexer realization is pass gate as well. Two-phase clocking is used in order to compare the DET-LM to the single-edge MSL and to draw some conclusions on the usefulness of the latch-mux arrangement as an alternative to the single-edge designs.

Circuit Operation   During the *Clk* high phase, the upper master latch is transparent and data are stored at the input of the second pass-gate stage. When *Clk* goes low, the stored upper master data updates the node Q. At that time the input pass gate of the upper master latch is turned off, disabling any further updates to the Q via the upper master path. Similarly, the lower master stage is transparent

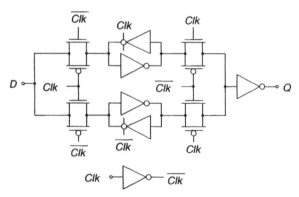

Figure 8.34. DET Latch-mux circuit. (Llopis and Sachdev (1996), Copyright © 1996 IEEE.

during *Clk* low. The upper and lower paths work in push-pull fashion, alternating the data flow from the upper to the lower path in each clock phase. It is worth noting that this particular implementation of the slave-mux carries a certain risk with it. Suppose that the capacitance at node *Q* is much bigger than that at the input to the slave-mux pass gate. In that case, when the pass gate is turned on by the $\overline{Clk}$ high for the upper path, charge sharing from *Q* back through the pass gate can flip the feedback in the master latch before the master's information is passed forward through the pass gate. This is of special concern in this type of design where the capacitance of *Q* is easily much bigger than at the output of the master, due to the parasitic capacitance of the mux pass gates and the output inverter.

8.5.2. DET C²MOS Latch-mux

DET-C²MOS-LM by Gago et al. (1993), Fig. 8.35, is a dual-edge version of the C²MOS MSL, Fig. 8.4. The latch design is conventional clocked CMOS, with some clock transistors shared by different stages. The multiplexer consists of two

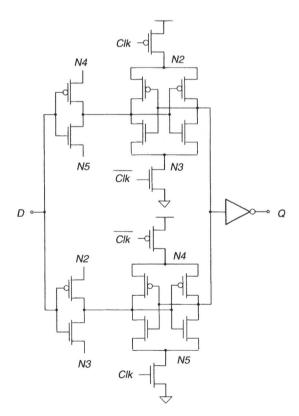

Figure 8.35. C²MOS Latch-mux. (Gago et al. 1993), Copyright © 1993 IEEE.

clocked CMOS inverters and the output buffer inverter. The second clock phase is generated locally.

Circuit Operation   During the *Clk* high phase, data are passed through the upper C^{2}MOS master gate with nodes *N*4, *N*5. At the same time, the latch in the lower path keeps the previous state and actively drives node *Q*. When *Clk* goes low, the upper latch becomes transparent and passes the data from the upper C^{2}MOS to *Q*. The upper and lower latches become opaque, while lower gate becomes transparent, enabling the data to be updated in the next clock phase.

The charge-sharing problem, present in DET-LM, is also present in this design. Consider the situation where initially *Clk* is high, *Q* is low, and the output of the upper latch is low. When *Clk* goes low, and $\overline{Clk}$ goes high, it is possible for *Q* to flip the input to the latch and not get updated, thereby loosing the information at the input of the latch. The feedback speed has to be carefully adjusted, optimizing the setup time of the latch and the overall CSE overhead. Another parasitic effect deserves to be mentioned. It is possible for data to feed through to the state node, *Q*, from the latch through the Miller capacitance of the tristated latch. For example, if the lower latch is transparent and the upper latch is opaque, the data can feed through the opaque latch via the Miller capacitance to induce noise on the *Q* state, which has to be absorbed by the lower transparent latch.

8.5.3. DET Pulsed-Latch

The PL has a very simple structure, consisting of the set of pass gates that define the transparency window, buffer inverters, and weak feedback path to keep the value stored in the PL output at the end of the transparency window. The clock delay line of four inverters defines the transparency window. There are two timing windows when the latch is transparent. The first is determined by the overlap of the clock, *Clk*, and the clock delayed by the three inverters, $\overline{Clk}_1$, and the second, is determined by the first and fourth delay of the clock, $\overline{Clk}$ and *Clk*$_2$ (Fig. 8.36b).

The original design by Strollo et al. (1999) is semistatic, that is, the feedback keeper was implemented for only the high level of the output, *Q*, and the pass-gate forward path was implemented using n-MOS transistors only. The design is modified by the addition of the complementary feedback and full transmission gate (Fig. 8.36). Without the modifications, the original structure exhibited a large delay for low-to-high transition, which caused the delay to be twice as long compared to any other design.

Circuit Operation   Let us first discuss the operation of the single-edge-triggered PL, Fig. 8.36a. At the rising edge of the *Clk*, both pass gates are turned on in the duration of three inverter delays — from *Clk* going low-to-high to $\overline{Clk}_1$ going high-to-low. At the same time, the feedback is disengaged, allowing for rapid propagation of the input signal to the output. After this transparency window of three inverter delays, $\overline{Clk}_1$, goes low and the first pass gate is turned

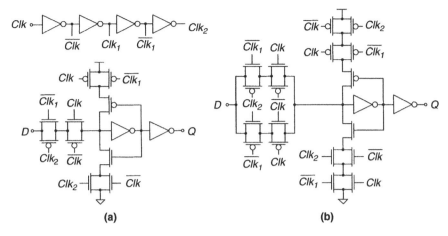

Figure 8.36. Pulsed-latch: (a) single-edge, (b) dual-edge triggered.

off, making the latch opaque, while the feedback is engaged to preserve the state of the latch.

The exact operation is a bit more complicated because the input data low has a transparency window of three inverter delays from Clk to $\overline{Clk}_1$, since the signal passes through n-MOS pass gates, while the input data high has a transparency window of three inverters from $\overline{Clk}$ to Clk_2, which is shifted in time by one inverter delay. This causes the setup and hold times to be different for the data high or low. The setup time is shorter and the hold time is longer for the input high case, since the pass gate turns off one inverter delay later for the input high.

The dual-edge counterpart is shown Fig. 8.36b, with the upper pass-gate pair transparent for three inverter delays after the rising edge of the Clk and the lower pass-gate pair transparent for three inverter delays after the falling edge of the Clk. As in the single-edge-triggered case, the position of the transparency window is data dependent, and the transparency window for the upper pass-gate pair is one inverter delay later for the D high than that for the D low. The situation is reversed for the lower pass-gate pair, where the transparency window occurs one inverter delay later for the D high than that for the D low.

8.5.4. DET Symmetric Pulse Generator Flip-Flop

The DET symmetric pulse generator flip-flop (DET-SPGFF) by Nedovic et al. (2002) is a novel flip-flop design, featuring a narrow data transparency window and clockless output multiplexing scheme. The circuit schematic is shown in Fig. 8.37. The first stage is symmetric and creates the data-conditioned clock pulse on each edge of the clock (at node S_X on the rising and S_Y on the falling edge of the clock). The second stage is a two-input NAND gate that effectively serves as a multiplexer, implicitly relying on the fact that nodes S_X and S_Y alternate in being precharged high, while the clock is low and high, respectively.

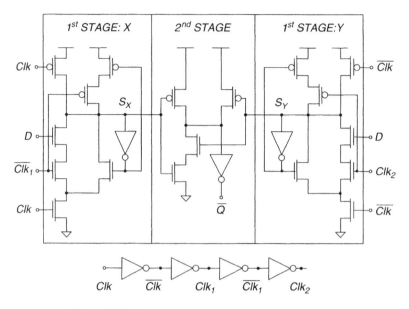

Figure 8.37. DET symmetric pulse generator flip-flop.

This type of output multiplexing is very convenient, because it does not require clock control. The clock energy is mainly dissipated for pulse generation in the first stage.

Circuit Operation   Let us examine the operation of the DET-SPGFF in more detail, considering the stage topology described in the previous section. Assume that the *Clk* is initially, low and *D*, $\overline{Q}$, S_X, and S_Y are high. When the *Clk* goes high, $\overline{Clk}_1$ is still high for the length of three inverter delays, enabling the discharge of node S_X since input *D* is high. As S_X goes low, it pulls $\overline{Q}$ low, since the second stage is a NAND gate with inputs S_X and S_Y. The event of S_X going low disables the pull-up path for S_X via the feedback inverter in the first stage, so S_X remains static low, even if *D* changes or the main signal path is disabled by $\overline{Clk}_1$ going low three inverter delays after the *Clk* went high. Node S_X is pulled down by the path enabled with the *Clk* and the feedback inverter. When the *Clk* goes low, node S_X is precharged, enabling the node S_Y to pass its value to the output $\overline{Q}$ in a similar fashion as the node S_X in the previous phase of the clock.

8.5.5. Comparison

The comparison of the DET-LM and corresponding MSL schematics reveals the behavior and performance relationship between MSLs and latch-muxes, in general. It is seen that the latch-mux has two equally critical paths that are somewhat shorter than that of the MSL (the delay of a multiplexer versus the delay of a latch in the second stage). As with the MSL, the activities of the internal nodes in the latch-mux are found to be directly proportional to the activity of the input. This indicates good power-consumption scaling with the activity, and preservation of the beneficial power-consumption features of the MSL. The advantage of the DET-C^{2}MOS-LM design at the circuit level compared to both DET-LM and single-edge C^{2}MOS is in the efficient multiplexer realization and sharing the clock transistors, which reduces overall clock load and clock power consumption. In contrast to latch-mux implementations, DET-PL requires that a large number of transistors be added and/or that the cell size be increased in order to obtain dual-edge functionality. This easily offsets the clock energy savings from halved clock frequency. The DET-SPGFF makes very fast operation possible with good power savings, yielding an overall best energy-delay product.

The delay comparison of single- and dual-edge-triggered devices is illustrated in Fig. 8.38. The advantage of latch-mux topologies is in the smart implementation of the latch-mux arrangements, DET-LM and DET-C^{2}MOS-LM. The PL structures are more complex, and their straightforward implementations are shown to increase both the delay and power. The DET-SPGFF benefits from both design approaches and results in the best delay.

Power consumption is compared in Fig. 8.39, at an average data activity of 50%, with single- and dual-edge clocks at 500 MHz and 250 MHz, respectively. The main conclusion is that although a potential for clock power savings exists due to the halved clock frequency, usually the latch capacitance switched by the clock is doubled to facilitate the multiplexing operation. This leaves the total CSE clock power roughly unchanged. With a smart design, it is possible to save some amount of the clock power dissipated inside the latch.

The energy-delay product, as an adequate measure of overall performance, is illustrated in Fig. 8.40. These data confirm that some savings are possible with a smart and efficient design of the latch-mux implementations, while pulsed latches

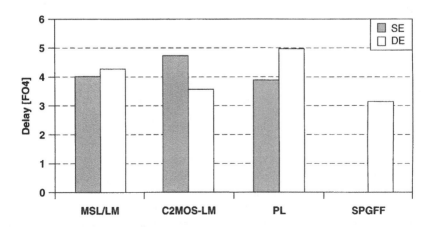

Figure 8.38. Delay comparison, SET vs. DET.

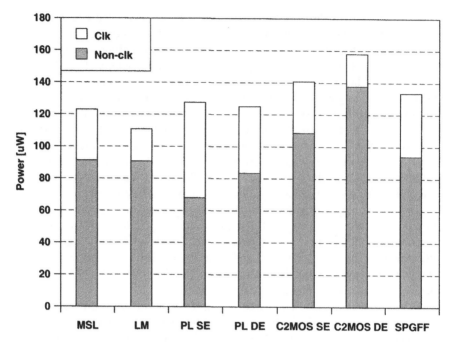

Figure 8.39. Power consumption comparison, SET vs. DET (0.18 μm, high load).

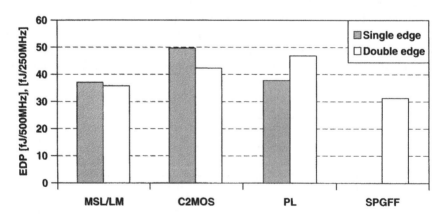

Figure 8.40. EDP comparison, single vs. dual-edge triggered clocks (0.18 μm, high load).

suffer from increased complexity and it is much harder for a designer to depart from their straightforward implementation.

Although the comparative analysis of energy dissipation in single-edge-triggered CSEs versus their dual-edge-triggered counterparts is interesting, it does not completely illustrate the benefit of using the dual-edge-triggered CSEs. It is indeed hard to design dual-edge-triggered structures that would consume less power (especially the clock power) than their single-edge-triggered counterparts. The true savings, however, are in the power consumed by the clock distribution network. The clock distribution example in Section 6.3 best illustrates the advantages of the halved clock frequency, despite the potential increase in total clock load, when dual-edge-triggered CSEs are used. The main conclusion is that the dual-edge-triggered design is always a better choice as long as it maintains the clock load capacitance at less than roughly twice that of the single-edge-triggered design.

8.6. SUMMARY

The choice of the CSE topology depends on the target application. CSE delay overhead is still the most dominant parameter in high-speed systems, although the energy consumption, especially the clocking component, is of increasing concern. Pulsed-latches and flip-flops offer the smallest data-to-output delays, due to negative setup time or the fast direct-path property, respectively. In addition to the small delay, these structures offer some degree of clock and data uncertainty absorption, which is of increasing importance in modern high-speed systems. This property, however, does not come for free and is traded for increased risk of hold time failure.

In low-energy applications, CSEs based on MSL pairs in general have several advantages over those based on PLs. MSLs tend to have better race immunity at the expense of increased delay. PLs and flip-flops generally dissipate more energy, and therefore practical applications where low energy is of primary concern would involve M-S topologies. MSLs are likewise preferred over PLs and flip-flops when performance is not the main design goal. It is important to note that not all of the parts of the processor are on the critical path, and low-energy, conservative MSLs can be used in these blocks to reduce the overall energy dissipation and alleviate global clock load requirements.

From the examples presented, we can summarize some of the main methods used in the design process to achieve low-energy consumption in clocked storage elements. The preceding examples have indicated that low-energy design can be accomplished to a certain degree in a systematic fashion, implying that the designer still has various degrees of freedom. Low-energy designs can be effectively concocted with methods that employ clock gating, reduced swing clocking, or dual-edge triggering. While the principles of operation of CSEs are different, the guidelines used to optimize them are essentially the same, differing only in the physical realization and in the resulting trade-offs made with regard to circuit delay and internal race immunity.

CHAPTER 9

MICROPROCESSOR EXAMPLES

The purpose of this chapter is to recapitulate the material presented in this book through various examples. It also presents the state of the art in microprocessor design from the standpoint of clocking and clocked storage elements. In this chapter we analyze the clocking techniques and clocked storage elements used in four leading microprocessor design houses: the Intel Corporation, Sun Microsystems Inc., the Digital Equipment Corporation (unfortunately, this latter company is not in existence any longer as of the writing of this book) and the IBM Corporation. We have chosen to emphasize different aspects each time we describe the techniques and designs used, and to go into depth on different solutions for clocking of high-performance and low-power systems.

In the section describing clocking techniques used in Intel® microprocessors we describe an active clock deskewing technique that was used in Pentium® processors. This discussion supplements the subject of clock distribution that was only briefly touched on in Chapter 1. We do not place much emphasis on the clocked storage elements used in Intel microprocessors in this section for two reasons: first, Intel never explicitly published the CSE topologies used in their microprocessors, and second, we have presented those CSE topologies known to us in the previous chapters.

The section describing techniques used by Sun Microsystems Inc. presents an overview of the development of the semidynamic flip-flop from its inception (described in patents filed by Sun Microsystems) to the final circuit used in their latest RISC processor, Ultra-Sparc-III®. This type of flip-flop represents one of the fastest single-ended flip-flops today and presents an interesting discussion on flip-flops with a "soft-edge" property that are capable of time borrowing and absorption of the clock uncertainties.

The section on Alpha® processors, developed by the Digital Equipment Corporation, describes various methods of microprocessor clocking. The Alpha processor is particularly interesting because it was a performance leader through the last decade of the last century. It is interesting not only from the point of view of the CSEs used, but in showing how this decision affected the clocking strategy used. The first generation of Alpha (WD21064) used single-clock latches, which are very similar to TSPC latches (Yuan and Svensson 1989). In the second-generation Alpha (WD21164), a pass-gate latch design was chosen. These latches also demonstrated the importance of incorporating logic into the CSE in order to accommodate the demand of shorter pipeline stages. The third generation Alpha (WD21264) used a sense-amplifier flip-flop (SAFF) (Madden and Bowhill 1990). Development of this flip-flop and its evolution into its final form (Oklobdzija and Stojanovic 2001; Nikolic and Oklobdzija 1999) were described in the previous chapters. This particular flip-flop is still among the fastest and most energy-efficient CSEs today. In order not to limit the performance of the processor, a hierarchical clock grid was introduced in the third and fourth generations of this processor.

Finally, in the section dedicated to the IBM Corporation, we describe design for testability techniques, specifically IBM LSSD, and IBM's particular practice of using latches and not flip-flops. We describe four recent microprocessors designed by IBM: the IBM S/390 G4; the experimental IBM PowerPC (the first one to break the 1-GHz barrier); a low-power champion PowerPC 603; and the IBM Power4. The interesting aspect of IBM designs is the ability to tune the *edges* of the clock, thus operating the processor in two modes: high-speed, and test and debug. The emphasis on diagnostic and machine bring-up is particularly important in a robust and high production-quality design. Various ways of incorporating logic and the scan function into the latch are of a particular interest.

This chapter brings together all the techniques described in this book and shows their relevance by describing the ways those techniques were incorporated into the most advanced microprocessors as of this writing.

9.1. CLOCKING FOR INTEL MICROPROCESSORS

Table 9.1 lists the key design parameters of the three generations of Intel microprocessors for desktop PCs (*source*: *Microprocessor Report Journal*, online: http://www.mdronline.com/mpr/). Consumer PCs based on the Pentium II processor featured new technologies such as DVD players and AGP graphics. The Pentium 4 was the first microprocessor to break the 2-GHz mark in clock speed. It features an increased number of pipeline stages, relative to its predecessors. Fabricated in 0.18 μm 6-metal-layer technology, it had almost double the number of transistors compared to Pentium III, and dissipated up to 67 W of power.

This section describes clock generation and distribution for some of the recent generations of Intel microprocessors. The emphasis of this section is on clock distribution and active deskewing circuits that have tight control of the clock

Table 9.1 Intel Microprocessor Features

	Pentium II	Pentium III	Pentium 4
MPR issue	June 1997	April 2000	Dec 2001
Clock speed	266 MHz	1 GHz	2 GHz
Pipeline stages	12/14	12/14	22/24
Transistors	7.5 M	24 M	42 M
Cache (I/D/L2)	16 K/16 K/—	16 K/16 K/256 K	12 K/8 K/256 K
Die size	203 mm^2	106 mm^2	217 mm^2
IC process	0.28 μm, 4 M	0.18 μm, 6 M	0.18 μm, 6 M
Max power	27 W	23 W	67 W

skew. An adaptive digital deskewing technique applied to the IA-32 Pentium Pro family is described first, followed by the clock generation and distribution in the first IA-64 microprocessor. The section concludes with the clocking scheme used in the Pentium 4 microprocessor. The examples and circuit diagrams presented in this section are adapted from the Intel papers presented at the International Solid-State Circuits Conference (ISSCC) over the last five years. The examples show detailed implementation of the clocking circuits reported in these papers.

9.1.1. IA-32 Pentium Pro

An adaptive digital deskewing technique is employed in the 450-MHz IA-32 Pentium Pro (P6 family) microprocessor (Schutz and Wallace 1998). The global skew for the clock distribution network in this 7.5 M transistor 0.25-μm technology microprocessor design is only 15 ps, down from more than 60 ps with the deskewing circuit inactive. The clock deskewing scheme used in the Pentium Pro is described in detail below, as presented by Geannopoulos and Dai (1998).

Clock skew is managed by the adaptive digital deskewing circuit. The deskewing circuit equalizes two clock distribution spines by compensating for delay mismatch in the left and right spines of the microprocessor clock network. The circuit is composed of delay lines in both spines, a phase detection circuit, and a controller, as illustrated in Fig. 9.1. The phase-detection circuit determines the phase relationship between the two spines and generates an output based on this phase relationship. The controller takes the phase-detection information and makes a discrete adjustment to one of the delay lines, minimizing the clock skew between the two spines. The main building blocks of the deskewing circuit are described next.

Figure 9.2. is a block diagram of the delay line and the delay shift register. The tunable digital delay line is implemented with two inverters in series, each loaded with a bank of eight capacitive loads. The content of the delay shift register determines the capacitive loads for the two inverter outputs. Both n-MOS and p-MOS transistors are used to make each of the capacitive loads, in order to reduce the signal slope and balance the rising and falling edge of the inverters.

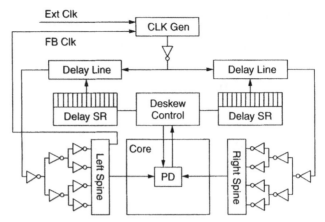

Figure 9.1. Clock distribution network with deskewing circuit. (Geannopoulos and Dai 1998), Copyright © 1998 IEEE.

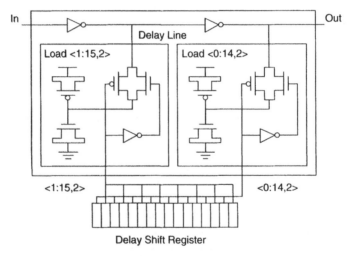

Figure 9.2. Delay shift register. (Geannopoulos and Dai 1998), Copyright © 1998 IEEE.

Capacitive loads are designed to allow 17 monotonic discrete steps of delay, with the average delay per step of 12 ps. The capacitive loads are added alternately to the two inverter outputs. The use of two inverters allows the load to be split between two drivers and also provides noninverting delay.

The phase-detection circuit is shown in Fig. 9.3. It consists of two symmetrical PDs and an adaptive noise-band filter. Each PD is designed with four $S-R$ latches in a pipelined configuration to reduce the probability of metastability propagating into the control logic as shown by Geannopoulos and Dai (1998).

Each PD has one of the sampled clock signals delayed by a controlled amount, Δn, of discrete time units. In PD1, the clock signal from the left spine is delayed

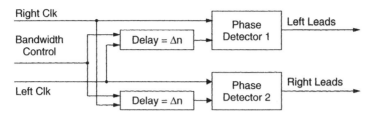

Figure 9.3. Phase detector. (Geannopoulos and Dai 1998), Copyright © 1998 IEEE.

by Δn, and then compared to the nondelayed clock signal from the right spine. Likewise, in PD2, the clock signal from the right spine is delayed by Δn and compared to the nondelayed clock signal from the left spine. Outputs of PDs indicate which of the clock signals arrives earlier. The possible combinations are: left is leading right, right is leading left, and both left and right are within the noise-band delay, $\pm \Delta n$, where the discrete time step was nominally 12 ps. Adaptive filtering reduces latency and allows the system to correct for AC power-supply-related components of skew variation.

The clock distribution network in a 7.5 M, 0.25-μm technology IA-32 P6 family microprocessor design (Schutz and Wallace 1998) has >60 ps of skew from left to right with the deskewing circuit inactive. With the deskewing circuit active, the skew was reduced to 15 ps. The digital deskewing circuit for clock distribution cancels out the skew (load, interconnect, and device mismatches). It also compensates for the *dynamic* variations of temperature and voltage gradients between the two spines during all phases of active microprocessor operation.

9.1.2. First IA-64 Microprocessor

The clock generation and distribution in the first IA-64 microprocessor is very much like that in the IA-32 Pentium Pro described in previous section. The IA-64 achieves a low skew through distributed programmable deskew units (Rusu and Tam 2000). The microprocessor is supplied by an external differential clock running at the system bus frequency. A PLL takes this clock and generates the high-frequency internal clock running twice as fast. The clock distribution architecture for IA-64 has three main components: (1) a balanced global clock tree, (2) multiple deskew buffers with balanced tree structures that drive the regional clock grids, and (3) multiple local clock buffers tapping these regional grids. In addition to the global clock, a separate reference clock is distributed along with the global clock to complete the deskew architecture. A block diagram of a clock distribution topology is shown in Fig. 9.4.

The global clock (the core clock and the reference clock) is distributed as a balanced H-tree. The core clock and the reference clock from the PLL are distributed to eight deskew clusters, each holding up to four deskew buffers that drive regional clock grids. Due to a high-frequency operation, the interconnect

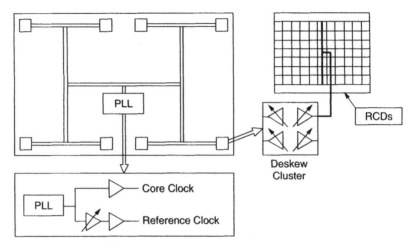

Figure 9.4. Clock distribution topology. (Rusu and Tam 2000), Copyright © 2000 IEEE.

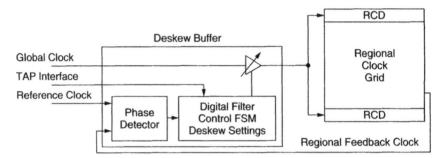

Figure 9.5. Deskew buffer architecture. (Rusu and Tam 2000), Copyright © 2000 IEEE.

model includes inductive effects. The placement of the intermediate buffers and the H-tree structure in order to provide minimal overall delay of the global clock is optimized according to a detailed *RLC* interconnect model. In addition, all the clock routing is fully shielded for the best noise immunity and good ground return paths. This special shielding and routing of the reference clock made it a better clock than the other clock signals, and it was therefore used as a reference.

The block diagram of a deskew buffer is shown in Fig. 9.5. It consists of a PD, a state machine, and a digitally controlled analog delay line. The PD compares the timing of the reference clock and a sampled feedback clock from the regional clock grid. A digital low-pass filter is used to eliminate variations from the phase comparison. The low-pass filter tracks the result of the last four PD comparisons and makes an adjustment if all of the last four measurements are identical. The delay of the digitally controlled analog delay line is adjusted in accordance with the phase-comparison results. A state machine controls deskew. The digitally

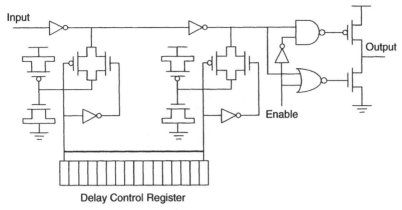

Figure 9.6. Digitally controlled delay line. (Rusu and Tam 2000), Copyright © 2000 IEEE.

controlled analog delay line shown in Fig. 9.6 supports 20 linear settings that cover a deskew range of 170 ps with an average step size of 8.5 ps.

Deskew is based upon a single reference clock that has the same delay relative to all the regional feedback clocks. This is achieved by inserting an average regional clock delay at the central reference clock generator. In the clocking topology the skew caused by distribution mismatches from the global clock to the regional clock, and the load mismatches at the regional clocks are replaced by the skew of the reference clock and the uncertainty of the PD. The skew of the reference network can be controlled due to its reduced span (about half the span of the normal clock), balanced topology, and fixed predictable loading (Rusu and Tam 2000).

The output of each deskew buffer is routed through a balanced tree to the distributed regional clock drivers, which drive a uniform clock grid that is used to achieve easy access from the underlying blocks. For power reduction, the clock grid is distributed only over active circuit areas. Each regional clock grid is independent. Independent regional clock grids make it possible for skew due to loading differences between the regions to be explicitly accounted for, because a single reference clock is used. If the design were implemented with only a single or a couple of grids, it would result in excessive skew (Geannopoulos and Dai 1998). Figure 9.7 plots the simulated clock skew for the worst-case region using extracted layout parasitics. The skew within a regional clock grid is less than 25 ps.

In this design, several types and sizes of local clock buffers are available as standard cells, including support for the delayed clocks used by the time-borrowing domino and clock gating for power reduction. The timing analysis tools model the delay of the local clock buffers, so designers can add skew to the local clocks as long as they meet the cycle-time and hold-time constraints (Rusu and Tam 2000). Figure 9.8 shows the experimental clock-skew measurement results. The worst skew of all regional feedback clocks between all deskew

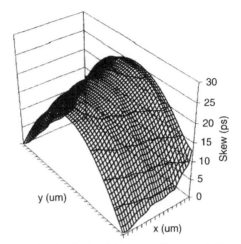

Figure 9.7. Simulated regional clock-grid skew. (Rusu and Tam 2000), Copyright © 2000 IEEE.

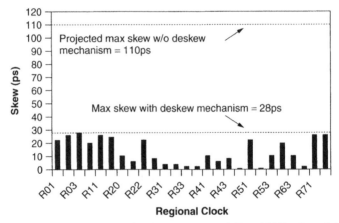

Figure 9.8. Measured regional clock skew. (Rusu and Tam 2000), Copyright © 2000 IEEE.

buffers is 28 ps. The equivalent skew without this deskew mechanism would have been more than 110 ps.

9.1.3. Pentium 4

The clock network for the Pentium 4 microprocessor is an example of a multi-GHz clock network. The clock network topology has three separate core and three input/output (I/O) bus frequencies, for a total of six clock frequencies running concurrently.

A PLL synthesizes the core and I/O clocks from a differential off-chip reference clock that is used to achieve maximum common-mode noise rejection. A two-stage double-differential clock receiver converts the low-swing differential clocks to a single-ended reference clock. The receiver is optimized to reduce input reference jitter due to signal, power supply, and temperature variations (Kurd et al. 2001).

In this multi-GHz design, the common, address, and data I/O busses operate at three different frequencies. The common clock bus operates at the same frequency as the system bus frequency; the address bus operates twice as fast; and the data bus operates at four times the system bus frequency. Figure 9.9 contains a block diagram of the core and I/O clock generation. The common and address clocks, and the feedback to the core PLL are generated using a programmable divider. The clock-enable divider generates enable signals to select the desired edges of the core clock. This allows logical verification of all clocks, avoids multiple global clock trees, and simplifies the interface with the core (Kurd et al. 2001).

In order to generate a centered strobe for all bus-to-core frequency ratios without compromising the outbound data timing margin, the data strobes that operate at four times the system bus frequency are generated by a separate PLL. The deskew synchronization state machine ensures sufficient setup/hold to account for any phase error and jitter. This comes at the expense of one cycle of latency for all the signals crossing the core-I/O clocking domains. The inbound data are latched using clocks derived from the received strobes. For proper data integrity,

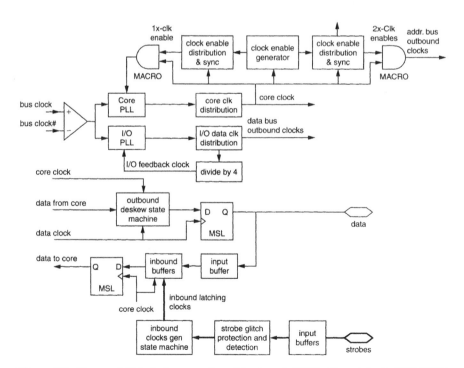

Figure 9.9. Core and I/O clock generation. (Kurd et al. 2001), Copyright © 2001 IEEE.

differential strobes with a glitch-protection/detection circuit are used. The filtered strobes clock the inbound clock-generator counter to generate clocks to latch the data in the eight deep inbound buffers. The data are then read out from these buffers using another counter that is clocked with the core clock (Kurd et al. 2001).

The global core clock distribution consists of a modified binary tree spanning multiple clock spines along the width of the die. The global distribution tree terminates in 47 domain buffers, producing 47 independent clock domains. Each domain buffer consists of a programmable delay stage controlled by a 5-b domain deskew register that determines the edge timing for the domain clock. The default value for the domain deskew register is loaded from a programmable fuse array at power-up, but can also be overridden though the test access port (TAP) for debug. This gives a convenient way to debug interdomain speed paths. A four-stage hierarchical network of phase detectors provides the means for comparing the rising edge clock timings of all domain clocks. Domain buffers can be disabled to power-down large functional units to save power (Kurd et al. 2001).

The clock repeaters in the global distribution network use an RC-filtered power supply to suppress clock jitter due to supply switching noise. The RC filtering provides 12-dB noise attenuation from the core power supply, reducing the cycle jitter by a factor of 4. Figure 9.10 is a logical diagram of the clock distribution network.

Systematic sources of skew (design convergence tolerances, layout mismatches, etc.) and random skew sources (within-die variation) are compensated by a static clock-deskewing scheme that employs the delay adjustment feature of

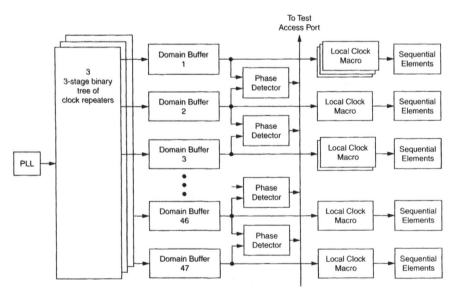

Figure 9.10. Logical diagram of core clock distribution. (Kurd et al. 2001), Copyright © 2001 IEEE.

the domain clock. An interdomain clock skew of <20 ps is achieved, and silicon data have shown up to 10% frequency improvement due to skew compensation. There is an additional flexibility to intentionally skew the domain clocks to maximize operating frequency. According to Kurd et al. (2001), up to one speed bin improvement was achieved in early silicon samples of the Pentium 4.

High microarchitectural performance is achieved by operating critical segments (e.g., ALU) of the Pentium 4 at double the core clock frequency. On the other hand, to conserve power, area and design effort, noncritical segments of the die operate at half the core frequency. The multiple clock frequencies are generated at the local clock macro level, without any additional clock skew penalty in the frequency-domain interface signals. The double-frequency clock is a pulsed clock created on both the rising and falling edges of the domain clock. The pulsed clock allows simple latches to be used as MSLs, reducing power and layout area, so the pulsed clock usage is extended to all frequency domains. To compensate for the extra inversion needed to provide high-going pulses from the falling edges of the clock compared to the rising edge of the clock, the duty cycle of the clock coming out of the PLL is designed so that the rising edge is one inversion delayed from the 50% duty cycle point. A special divide-by-two circuit produces the non-50% duty cycles of the clock from the VCO output of the PLL. The pulse widths of the pulsed clocks can be modified through control register setting. Figure 9.11 shows the logical implementation of the different types of local clock macros.

In addition to the ability to use the clock compensation to support timing debug, on-die clock stretch/shrink (ODCS), duty cycle adjust, and bypass modes are supported. In ODCS, the clock/duty cycle injected into the network is manipulated deterministically to uncover speed path problems (Rusu and Tam 2000). The data-clock duty cycle is adjusted in a similar way at the I/O PLL to stress out the timing between sending and sampling data at the receiving agent. The bypass mode provides the ability to inject arbitrary clock waveform directly to the core, bypassing the PLL. The skew-measure circuit measures the phase difference between the feedback clocks to ensure that the skew and jitter between the two clocking domains are within the tolerance designed in the outbound deskew.

With ever increasing clock speeds and microprocessor die size, balancing the clock skew in large designs using simple RC trees is becoming less effective. The increased die size often times results in the insertion delay of the clock network of about 7−8 FO4 inverters, comparable to the clock period. In addition, due to process, voltage, and temperature (PVT) variations across large dies, clock skew is becoming a larger portion of the useful clock period. Another important issue associated with the gigahertz frequencies is the inductive effects, where a simple RC model is not valid anymore and should be replaced with a more accurate RLC model, due to the increased importance of parasitic inductance at high frequencies. Controlling the clock skew using simple RC-based methods is therefore not effective anymore. The active deskewing circuits used in the clock distribution in Intel microprocessors are a good solution to the increasing skew problem,

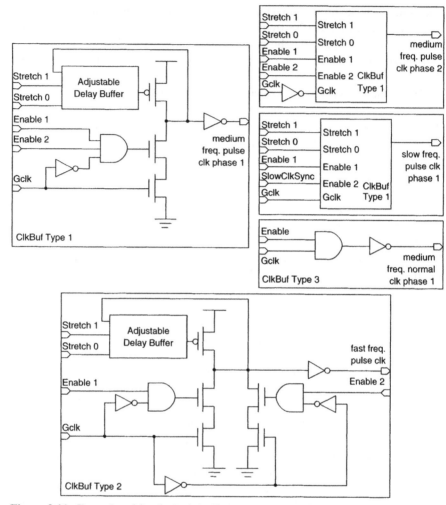

Figure 9.11. Example of local clock buffers generating various frequency, phase, and types of clocks. (Kurd et al. 2001), Copyright © 2001 IEEE.

because the deskewing circuits reduce the effects of PVT variations and parasitic inductance by actively tracking the temporal variations of these parameters.

9.2. SUN MICROSYSTEMS ULTRASPARC-III CLOCKING

Sun Microsystems has delivered three generations of high-performance Ultra-SPARC microprocessors over several years (Charnas et al. 1995; Lev et al. 1995; Greenhill et al. 1997; Lauterbach et al. 2000; Heald et al. 2000b, c). UltraSPARC microprocessors are based on 64-b SPARC V9 architecture extension of 32-b

Table 9.2 UltraSPARC Family Characteristics

	UltraSPARC-I	UltraSPARC-II	UltraSPARC-III
Year	1995	1997	2000
Architecture	SPARC V9, 4-issue	SPARC V9, 4-issue	SPARC V9, 4-issue
Die size	17.7×17.8 mm^2	12.5×12.5 mm^2	15×15.5 mm^2
Number of transistors	5.2 M	5.4 M	23 M
Clock frequency	167 MHz	330 MHz	1 GHz
Supply voltage	3.3 V	2.5 V	1.6 V
Process	0.5-μm CMOS	0.35-μm CMOS	0.15-μm CMOS
Metal layers	4 (Al)	5 (Al)	7 (Al)
Power consumption	<30 W	<30 W	<80 W

RISC instruction set. They are designed for process scalability while maintaining the main architectural features and backward compatibility. An overview of the characteristics for UltraSPARC family is given in Table 9.2.

Due to the reduction in the number of logic levels in the pipeline stage, tripled clock frequency with respect to its predecessor, and the increased impact of the clock uncertainties, clocking has become a major issue in the design of the UltraSPARC-III. This section reviews the system-level and circuit-level challenges and solutions applied in this processor.

9.2.1. Clocking

The targeted high clock frequency of the UltraSPARC-III (600 MHz originally, 1 GHz reported in Heald et al. (2000b)) requires high-quality, low-uncertainty clock generation and distribution. A dual-loop PLL (Bhagwan and Rogers 1997) is used to generate a high-frequency on-chip global clock from an external reference. The PLL is capable of switching from 1/32 to 1 of its VCO frequency, allowing for low-power and full functionality in the standby operation. Measured PLL jitter was 62 ps peak-to-peak.

In order to minimize clock skew, the global clock is distributed using a balanced clock network (tree), and then terminated by a global metal grid that serves as an equalizer for the arrival times of the clock signals that drive each major block (domain) on the chip. The global grid is locally buffered in order to achieve the uniform grid loading. The local buffers are also used for purposes of testability. The second level of the clock distribution is the local grid within each of the blocks, which is terminated by the clock buffers.

The large number of clock terminals (nearly 80,000 storage elements) imposes a large nonuniform load on the clock distribution network. In addition, the aggressive dimension scaling (the wires are taller than they are wide) gives rise to the crosstalk-injected noise. This is why the clock tree and the clock grid metal lines are shielded and their dimensions kept as uniform as possible.

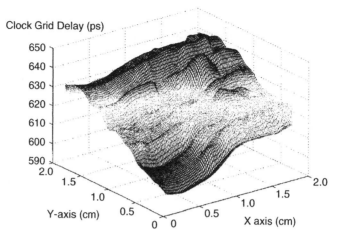

Figure 9.12. Clock distribution delay in UltraSPARC-III. (Heald et al. 2000a), Copyright © 2000 IEEE.

Simulated clock grid delay, or the portion of the total skew caused by imbalanced loads and path lengths, is shown in Fig. 9.12 as a function of the location on the chip. The total clock skew, which also stems from the supply voltage, temperature, and the process variations, was 80 ps (Heald et al. 2000b).

9.2.2. Storage Elements

With the aggressive circuit design applied in UltraSPARC-III in order to meet the targeted clock frequency, the number of logic levels per pipeline stage was reduced to eight. This increases the relative clocking overhead to the clock cycle, and the storage-element design becomes critical. UltraSPARC-III design uses the fast edge-triggered flip-flop family (Klass et al. 1999), with either static or dynamic consecutive logic driven by both static and monotonic dynamic output.

The basic flip-flop, SDFF (Klass 1998; Klass et al. 1999), is shown in Fig. 9.13. It consists of two functional stages: the first stage is a dynamic evaluation stage, and the second stage is a dynamic-to-static latch, found in the TSPC MSL (Yuan and Svensson 1989). The operation of SDFF is based on the local generation of an implicit clock pulse, first introduced in Partovi et al. (1996). There is a short period of time (transparency window) following the leading edge of the clock, during which a change in the state of the flip-flop is allowed. This approach allows fast switching of the flip-flop, since the critical path is short compared to MSLs, which makes it more suitable for high-speed applications. However, the hold time, determined by the moment the transparency window closes, is long. In addition, due to the high switching activity of the internal signals, the power consumption is large.

When the clock is at the low level, the node $\overline{S}$ is precharged high (transistor M_{P1} is on). The output level is maintained by the back-to-back inverters Inv_5 and Inv_6. The flip-flop is transparent during the short time window after the leading

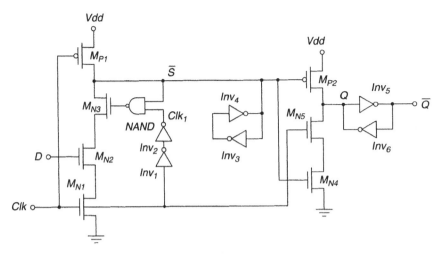

Figure 9.13. Semidynamic flip-flop. (Klass 1998), Copyright © 1998 IEEE.

edge of the clock, which is determined by the delay through inverters Inv_1 and Inv_2 and the NAND gate. If input D is high during that period, internal signal $\overline{S}$ evaluates to low (transistors M_{N1}, M_{N2}, and M_{N3} in Fig. 9.13 are on), which turns transistor M_{P2} on and pulls the node Q to the high level. If the flip-flop input D is low during this transparency window, the node $\overline{S}$ remains at the high level, and there will be no other opportunity for $\overline{S}$ to fall until the next clock edge. This high level on internal node $\overline{S}$ is used to force the node Q low via transistors M_{N4} and M_{N5}. After the transparency window has elapsed, the level of the node $\overline{S}$ is maintained by the back-to-back inverters, Inv_3 and Inv_4.

The SDFF employs a NAND gate in the positive feedback of the first stage (conditional shutoff mechanism). As soon as the first stage of the flip-flop starts evaluating and node $\overline{S}$ discharging, the output of the NAND gate is held at the high level, regardless of the state of its other input. This keeps the transparency window open even if the delayed clock Clk_1 switches to the high level. Consequently, the first stage of the flip-flop is able to evaluate for later low-to-high data arrival as opposed to the case when an inverter replaces the NAND gate. Therefore, the conditional shutoff mechanism improves low-to-high setup time. In the case where low-to-high data arrival is the only one expected to happen (e.g., dynamic logic drives the flip-flop), this mechanism can greatly improve performance. For static signaling, where both input transitions can occur during the evaluation, high-to-low setup time may become critical and the benefit from this positive feedback may not be seen.

The absence of the evaluation of node $\overline{S}$ causes node Q to reset, which is much faster than the time-critical low-to-high input transition. However, due to nonzero evaluation time through the first stage, a static-one hazard exists that manifests as a short glitch at node Q after the leading edge of the clock when both the preceding and following states of the flip-flop are high. The glitch increases the flip-flop's power consumption, reduces the noise immunity, and may corrupt the evaluation of the consecutive logic. This glitch, which is also seen in some other flip-flop designs (Partovi et al. 1996), is a disadvantage of the SDFF, and its propagation must be inhibited by transistor sizing.

Stripping the first stage from the standard static CMOS implementation and making it like dynamic logic styles, as done with the SDFF, effectively speeds up the response and allows for a simpler second-stage realization. This solution also eliminates the need for the transparency window in the second stage used in Partovi et al. (1996), and thus avoids disadvantageous asymmetry between the high-to-low and low-to-high setup times. However, both stages of the flip-flop can be in high impedance for up to half of the clock period. In order to improve the signal integrity, both nodes $\bar{S}$ and Q are made static by the back-to-back inverters (Inv_3–Inv_4 and Inv_5–Inv_6). This provides noise immunity similar to that of the domino logic gate, but results in the contention at the two nodes of the flip-flop and somewhat increases both delay and power.

Because of the remarkably small number of logic levels per pipeline stage (eight), an important property of the flip-flop family designed for UltraSPARC-III is the ease of logic embedding. The logic can be embedded in the flip-flop in Fig. 9.13 by replacing the transistor gated by the input D with an n-MOS network performing an arbitrary noninverting logic function, similar to the way it is done in domino logic style (Fig. 9.14). The example in Fig. 9.14b shows the embedding of the two-input XOR logic function into the flip-flop, that is, the value of node Q is functionally the same as if the output of a stand-alone two-input XOR gate is fed to the flip-flop. This allows a portion of the clocking timing overhead to be masked by the useful work performed by the embedded logic.

The dynamic versions of the basic flip-flop are shown in Fig. 9.15. The dynamic flip-flop is designed to drive the dynamic logic. It differs from a common domino gate by the shutoff transistors (M_{N3} in Fig. 9.15a; M_{N3} and M_{N5} in Fig. 9.15b) that allow evaluation only immediately after the leading edge of the clock. As with dynamic gates, its output precharges to the low level when the clock is low. Note that the shutoff of the differential dynamic flip-flop from Fig. 9.15b is delayed by the propagation delay inverters Inv_1–Inv_2 or Inv_3–Inv_4. This delayed shutoff decreases the flip-flop's setup time and increases its hold time. In effect, the timing overhead of the flip-flop in the long path is reduced at the expense of the increased short-path hazard.

The final version of the flip-flop used in UltraSPARC-III (Heald et al. 2000b) is shown in Fig. 9.16. Its principle of operation is very similar to that of the basic SDFF. It is modified, however, to use conditional keepers instead of back-to-back inverters in both stages. This modification is meant to reduce the impact of the energetic alpha particles from solder bumps, on the correct operation of the flip-flop. It is found that energetic alpha particles are capable of corrupting the levels of lightly loaded nodes that are not strongly driven at all times [less than 100 fC of the charge and less than 5 mA of the driving current (Heald et al. 2000b)]. The alternative solution to the soft-error problem is to increase the size of the back-to-back inverters in Fig. 9.13, which would seriously impair the performance or

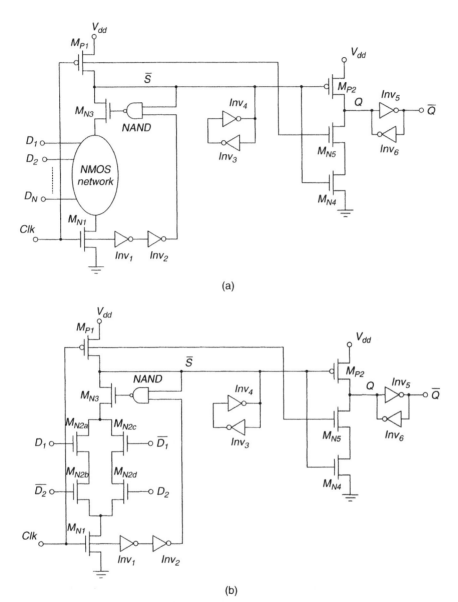

Figure 9.14. (a) Logic embedding in a semidynamic flip-flop; (b) two-input XOR function. (Klass 1998), Copyright © 1998 IEEE.

even functionality of the flip-flop due to the contention. In order to achieve the required robustness to the soft errors not compromising the performance, the sensitive nodes of the flip-flop are kept (restored) only when they are not driven otherwise. The low level at node $\overline{S}$ is restored only when the input clock is at the high level. The high level at node $\overline{S}$ is restored only when the conditional shutoff

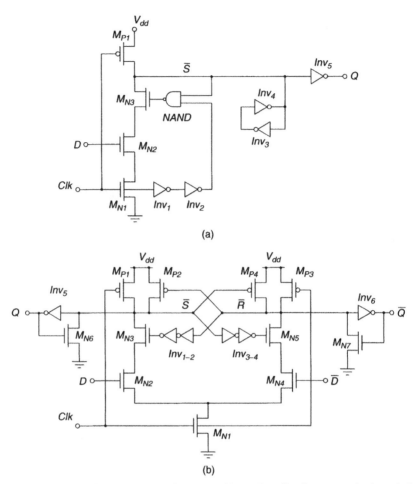

Figure 9.15. Dynamic versions of (a) semidynamic flip-flop: (a) single-ended; (b) differential. (Klass 1998), Copyright © 1998 IEEE.

transistor is off or the input D is low. The low level at node Q is restored only when node $\overline{S}$ is at the high level. The high level at node Q is restored only when the clock is low. In this way, flip-flop implementation is moved from the domino-like to static CMOS-like. Note that the static implementation of the flip-flop, driven by the signal integrity requirement, is more like the systematically derived flip-flop from Fig. 2.16 than it is to the original SDFF. The highly desirable logic embedding property is somewhat degraded compared to the basic SDFF, since the dual network of the smaller p-MOS keeper transistors needs to be implemented in addition to the n-MOS logic network of Fig. 9.14.

Clocking for the UltraSPARC-III microprocessor faced the complex combination of the design challenges due to the technology, large die size, and

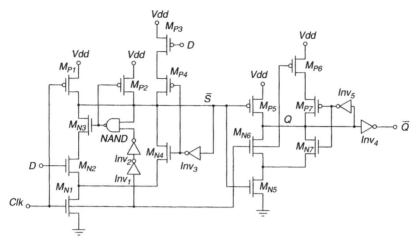

Figure 9.16. UltraSPARC-III flip-flop. (Heald et al. 2000a), Copyright © 2000 IEEE.

performance requirement. The performance-driven, high-power clock distribution system implements advanced methods in order to supply a high-quality clock to the large number of storage elements. The high-performance clocked storage elements are developed as an integrated part of the UltraSPARC-III microprocessor's clocking subsystem. A small number of logic gates per pipeline stage and an increase in the clock uncertainty make the performance of the UltraSPARC-III flip-flop a critical design criterion. The high speed and good logic-embedding property of this flip-flop allow the increase in the clock frequency and improve the testability of the design. However, the long hold time requires the use of advanced tools to identify and fix the fast-path violations. In addition, the large switching current that is caused by the high operating frequency and large number of transistors on the chip, together with technology scaling, draw attention to the issue of noise robustness in clocking.

The future of UltraSPARC architecture depends to a large extent on the scalability of its clocking subsystem. As the power consumption approaches the practical limits of heat removal and the number of transistors on the die increases, the clock has to adapt to the system of conditioned, globally asynchronous clock domains. The signal integrity issues, seen to be a problem in this UltraSPARC generation, can only become worse as the feature decreases, the transistor leakage grows, and the switching current increases. Thus, in order to continue performance scaling, the circuit design of future UltraSPARC microprocessors may need to be responsive to the power-saving and noise-robustness requirements, while retaining its high-speed operation.

9.3. ALPHA CLOCKING: A HISTORICAL OVERVIEW

In the past eight years, Digital has delivered four generations of high performance Alpha microprocessors, each by itself leading the state of the art of its time.

This has been achieved through process advancements, architectural innovations, and aggressive circuit-design techniques. This chapter gives an overview of the evolution of clocking techniques through the example of clock distribution and latch-design methodology in four generations of Alpha microprocessors, 21064–21364. The material presented is largely based on an excellent overview of Alpha microprocessor design by Gronowski et al. (1998).

Table 9.3 illustrates the key design parameters of the four Alpha microprocessor generations. The 21064 was the first implementation of the Alpha architecture. Designed to operate at 200 MHz in a 0.75-μm n-well CMOS process, it allowed about 16 FO3 gate delays per cycle, including latching, with a power dissipation of 30 W from a 3.3-V supply (Dobberpuhl et al. 1992). The die contains 1.68 million transistors, half of which represent noncache logic. The second generation, 21164, was designed to operate at 300 MHz in a 0.5-μm n-well CMOS process, with the number of FO3 gate delays reduced from 16 to 14 to enable a cycle time reduction of 10% beyond process scaling (Bowhill et al. 1995). With 2.5 million noncache transistors out of a total of 9.3 million, this processor dissipated 50 W from a 3.3 V supply. The 21264 was designed as the third generation in a 0.35-μm n-well CMOS process with a target speed of 600 MHz and the number of FO3 gate delays further reduced to 12, which provided an additional 10% clock-cycle reduction relative to the previous design (Gieseke et al. 1997). A nominal power supply of 2.2 V limits the power consumption to 72 W. The total number of transistors is 15.2 million, with a noncache transistor count of more than double that of the 21164.

The latest generation of Alpha microprocessor, 21364, contains the 21264 in its core, surrounded by level-two cache, a router unit, and a Rambus memory controller (Jain et al. 2001). It was designed to operate at a clock frequency of 1.2 GHz, in a 0.18-μm bulk CMOS process, dissipating 125 W from the 1.5-V supply. The total number of transistors is 152 million.

9.3.1. Clocking

The motto of the Alpha clocking system design can be stated as: "The primary objective of the clock system is to not limit the performance of the microprocessor"

Table 9.3 Alpha Microprocessor Features

	21064	21164	21264	21364
Number of # transistors [M]	1.68	9.3	15.2	152
Die size [mm^2]	16.8 × 13.9	18.1 × 16.5	16.7 × 18.8	21.1 × 18.8
Process	0.75 μm	0.5 μm	0.35 μm	0.18 μm
Supply [V]	3.3	3.3	2.2	1.5
Power [W]	30	50	72	125
Clk. frequency [MHz]	200	300	600	1200
Gates/cycle	16	14	12	12

(Gronowski et al. 1998). Indeed, the targeted operation frequencies required the generation and distribution of a very high-quality clock with very low skew, and the use of low-latency latches. Power supply noise, process variation, and interconnect delay introduce uncertainty in the timing of clock edges, reducing the maximum clock frequency. Moreover, slow clock edges cause uncertainty in latch timing and a possible hold time violation due to race-through.

The 21064 microprocessor departed from the traditional four-phase clocking style used in VAX machines. The choice to use two-phase single wire level sensitive clocking eliminated the dead time between the phases, resulting in a saving in overall cycle time. The robustness of the four-phase clocking scheme to race-through was maintained by careful selection of latch structures. The clock distribution network, based on a metal 2–metal 3 grid, and driven from the center (Fig. 9.17a), averages out the delays over different locations on the die. The plot of the clock skew across the die is shown in Fig. 9.18, with the largest skew of 240 ps, or equivalently, 0.8 FO3 gate delays, in the corner of the grid.

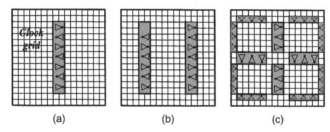

(a)   (b)   (c)

Figure 9.17. Alpha microprocessor final clock driver location: (a) 21064, (b) 21164, (c) 21264.

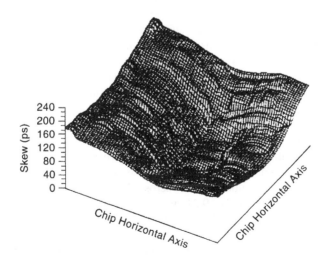

Figure 9.18. 21064 clock skew. (Gronowski et al. 1998), Copyright © 1998 IEEE.

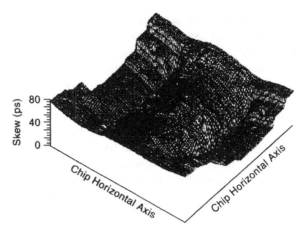

Figure 9.19. 21164 clock skew. (Gronowski et al. 1998), Copyright © 1998 IEEE.

By splitting the final driver into two banks placed midway between the center of the die and the edges (Fig. 9.17b), clock skew was halved in the 21164 microprocessor relative to the 21064, and uneven thermal distribution was avoided. A predriver is located in the middle of the die, distributing the clock to two driver banks. The plot of the clock skew across the die is shown in Fig. 9.19, with a top skew of 80 ps, that is, 0.4 FO3 gate delays. The plot clearly indicates the position of the two drivers, with skew increasing horizontally toward the ends and the middle of the die.

With the increase in microprocessor complexity, power consumption became one of the critical factors driving design decisions. Clock distribution was identified as one of the main components of high power consumption. Grid-based clock distribution networks introduce extra capacitance, leading to suboptimal power conservation. Hence, the 21264 clock distribution style departed from the global grid-based design, introducing a trade-off between the buffered-tree design, with its lower power but greater mismatch, and the grid-based design, with its higher power dissipation but lower mismatch.

For the first time, a hierarchy of clocks has been introduced, as shown in Fig. 9.20, that enable clock conditioning to save power and local clock manipulation to increase performance in critical sections, for example, using "time borrowing."

The global clock network is distributed in window-like configuration (Fig. 9.17c), with four grids driven by clock drivers from all sides to minimize the skew. A combination of H and X trees is used for the predriver to distribute the clock to the main clock drivers across the die (Friedman 1995). The global clock skew is shown in Fig. 9.21, with a maximum skew of 72 ps, that is, 0.5 FO3 gate delays. The plot clearly outlines the four grids, with skew peaking in the middle of each grid.

The fourth generation of Alpha microprocessor, 21364, introduced new challenges in the design of the clock distribution network. A plain extension of the

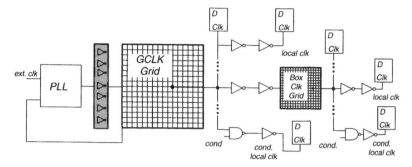

Figure 9.20. 21264 clock hierarchy. (Gronowski et al. 1998), Copyright © 1998 IEEE.

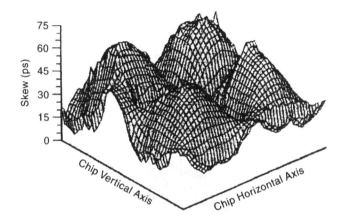

Figure 9.21. 21264 global clock skew. (Gronowski et al. 1998), Copyright © 1998 IEEE.

clock distribution technique from the core 21264 would not work over additional clock domains, such as the L2 cache clocks and the network-interface clock (*NCLK*), because of the enormous size of the die and projected power consumption of a single global reference clock. Four clock domains were created instead, as shown in Fig. 9.22, where *GCLK* is the global clock of the 21264 core, and *L2Lclk*, *L2Rclk*, *NCLK*, are synchronized with the *GCLK* using three *DLLs*, which reside at the root of each additional corresponding clock domain (Xanthopoulos et al. 2001).

The main role of a *DLL* is to "hide" the skew of the global clock grid from the center to the periphery from which the *NCLK* and other clocks would need to be distributed further. Using *DLLs*, the roots of the *NCLK* and other additional domains are referenced to the unskewed *GCLK* reference. In this way, all four domains are globally synchronized and the only skew that remains is the skew of each clock's local distribution network. Thus, both the skew and jitter are reduced, although any jitter introduced by the reference clock is directly passed to the domain clock through the *DLL*.

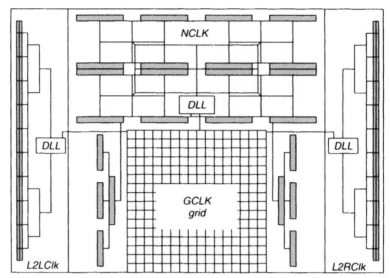

Figure 9.22. 21364 major clock domains. (Xanthopoulos et al. 2001), Copyright © 2001 IEEE.

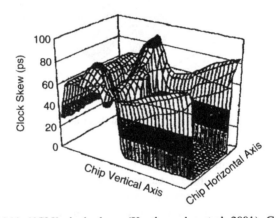

Figure 9.23. 21364, *NCLK* clock skew, (Xanthopoulos et al. 2001), Copyright © 2001 IEEE.

The design of the clock distribution network for *NCLK* domain was especially hard, due to the π-shape of the domain. Rectangular X trees are used to distribute the *NCLK* to the grid at the north of the core, and partial H trees are used along the sides. The plot of the *NCLK* skew performance only is shown in Fig. 9.23.

9.3.2. Clocked Storage Elements

Clock distribution and CSEs make up the backbone of every microprocessor system. They cannot be designed independently of one another, nor can they disregard the architectural features of the system. The preceding section discussed the evolution, or sometimes, revolution, in the clock design of four generations of the Alpha microprocessor. No specifics were given about the CSE design methodology. However, it should be noted right away that these two features are very tightly coupled and were designed concurrently, reflecting a global clock and CSE design methodology.

In each generation of Alpha microprocessor, new system requirements forced the changes in both clock distribution and latch circuit design and methodology. In addition to speed, other important goals have been minimal clock loading, low power dissipation, and small setup and hold times, that is, a narrow sampling window. In order to achieve improvement in speed beyond the process scaling, from generation to generation, one of the options, heavily used in Alpha design, has been to use a smaller number of gates in the pipeline. Starting with 16 FO3 gate delays per cycle in the 21064 (Table 9.2), and ending with 12 FO3 gate delays in the 21264 and 21364, the latch overhead became increasingly important. The need for short latency and setup time, as well as a free logic function with the inclusion of logic in the input or output stages of the CSE, became significant factors driving CSE design methodology.

The 21064's revolutionary two-phase, level-sensitive, single-wire clocking scheme, a break from the traditional four-phase scheme, required new design strategies (Dobberpuhl et al. 1992). These focused mainly on the more careful design of CSEs and was particularly careful to minimize the risk of race-through, which was not present in previous versions of Digital's microprocessors. The TSPC level-sensitive latches designed by Yuan and Svensson (1989), were the first CSEs to use a two-phase, single-wire clock. A variation of these latches was used in the 21064 to prevent data race-through. The latches used the unbuffered global clock directly, largely enhancing race immunity.

To understand the operation of the TSPC latches, refer to Fig. 9.24a. When Clk is high, P_1, N_3, and N_1 function as an inverter, complementing the D to produce X. Transistors P_2, N_4, and N_2 function as a second inverter, inverting X

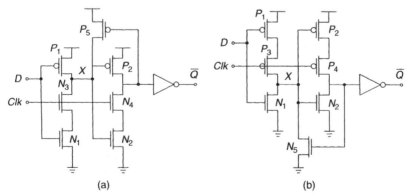

(a)   (b)

Figure 9.24. 21064 modified TSPC latches. (Gronowski et al. 1998), Copyright © 1998 IEEE.

to the output. The output of the second inverter is dynamic, and hence shielded by additional inverter stage, which increases immunity to coupling noise. When *Clk* goes low, two gated inverters are tristated to ground by N_3 and N_4 being turned off. Now, if D, X, and Q are initially high, low, and high, respectively, when *Clk* is low, the transition of D falling charges X to high, turning off P_2 and tristating Q from both power and ground. In the opposite situation, when D, X, and Q are initially low, high, and low, respectively, the transition of D rising tristates node X to high, leaving Q tristated to low. In summary, after Clk goes low, additional transitions on D leave nodes X tristated or driven high, and Q tristated to its initial value. This behavior is exactly that of the level-sensitive latch, which is transparent when *Clk* is high and opaque when *Clk* is low. The operation of the structure in Fig. 9.24b is dual to that in Fig. 9.24a. In the original structures, the only node exhibiting the unusual noise immunity risk is node X. This is because X can be tristated high, with Q tristated low when the latch is opaque (Fig. 9.24a), which translates into a dynamic node driving a dynamic gate that is very sensitive to leakage through P_2 charging node Q and destroying the data. To increase the noise margin at node X, (a) weak feedback device, P_5, was added to prevent X from being tristated high. The device should be sized to absorb any reasonable noise and keep P_2 turned off. Transistor N_5 plays an analogous role in Fig. 9.24b. The latches shown are just examples of the variety of latches used in the 21064, with embedded logic as gated AND and NAND, OR, and NOR gates. The zero-delay goal between the latches (as in the shift registers), their variety, hence different latency, setup and hold times, and clock uncertainty increased the risk of race-through as a major functional concern. This was addressed by paying special attention to the clock distribution and extensive latch simulations. Clock skew is functionally harmless if data propagate in the opposite direction to the clock waveform. In this case, no hold time violation is possible, but setup time can be violated, since the skew is subtracted from effective cycle time. In the case where both the clock and the data propagate in the same direction, clock skew can potentially cause a hold time violation, and hence, race-through. Since data progate from the periphery to the center of the chip, the radial distribution of the clock from the center of the chip prevents the data from overtaking the clock. Latch simulations involved exploration of process corners and parameters that could potentially cause the mix of any of the two latches to fail functionally. With 1.0-ns (3.3FO3 delays) *Clk* rise and fall times, latches showed signs of failure (Dobberpuhl et al. 1992).

Progressing to higher clock frequencies, the TSPC latch overhead became prohibitive for the 21164 design, hence a family of dynamic, level-sensitive, pass-transistor latches was used to minimize the latency of the latch. Clocking style remained single-wire, two-phase, requiring the use of phase A and phase B latches (Bowhill et al. 1995). Figure 9.25 shows A-and B-type latches, while Fig. 9.26 shows the embedding of logic in the 21064 and 21164 latch families.

Latch overhead reduction by using the embedded logic became very important, as the number of gate delays per cycle was reduced from the previous generation in order to increase the cycle time beyond the process scaling. Using

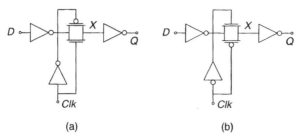

Figure 9.25. 21164: (a) phase-A latch, (b) phase-B latch. (Gronowski et al. 1998), Copyright © 1998 IEEE.

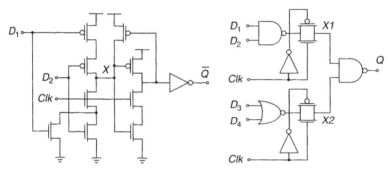

Figure 9.26. Embedding of logic into a latch: (a) 21064 TSPC latch, one level of logic, (b) 21164 latch, two levels of logic. (Gronowski et al. 1998), Copyright © 1998 IEEE.

this approach, the latching overhead was reduced to only one pass-gate delay in critical paths with structures, as in Fig. 9.26b. It is very important to note, however, that special care had to be taken to prevent the output node from coupling back onto the dynamic nodes (X) through the output transistors. The methodology required that inputs to the final logic gate come from the same latch type, phase A or phase B, in order to prevent the back-gate coupling effect. This effect occurs when the top transistor in a NAND gate n-stack is driven by a dynamic node, while the input to the bottom transistor of n-stack rises, pulling the intermediate node of the n-stack from $V_{DD}-V_t$ to ground. In that event, the dynamic node driving the top of the stack is pulled down as well via capacitance, C_{gs}. Coupling the output of the final gate to the dynamic nodes via Miller capacitance of the output transistors was reduced by the requirement that the latch nodes drive only the final two-input logic gate, using minimal routing.

Local generation of the second phase of the clock introduced one gate delay between the source latch becoming transparent and the destination latch becoming opaque, thus enabling the race-through and making the zero-delay requirement between the latches impossible to achieve. The race-through was prevented by controlling the skew on the globally distributed clock, precise sizing of the local clock buffer inside the latch, and requiring a of minimum of one logic delay element between all latches.

Driven by opposing requirements to increase the clock frequency by more than the process scale factor and to reduce the clock power, the design of the clocking strategy for the 21264 presented new challenges and resulted in global changes in clock distribution and latch methodology. To save power, conditional clocks were used mandating the use of slow, static latches. Hence, a family of flip-flops was used, based on the dynamic flip-flop shown in Fig. 9.27 (Matsui et al. 1994).

When employing a fast and sensitive regenerative sense-amplifier stage (Madden and Bowhill 1990) as a pulse generator, this structure has unnecessary overhead of two gate delays introduced by the cross-coupled NAND-based $S-R$ latch at the output. In critical paths, the static $S-R$ latch is replaced by the dynamic $S-R$ latch structure (Gieseke et al. 1991).

The use of flip-flops simplified the timing and race-through design issues that were magnified by the use of conditional clocks, but has also introduced new timing analysis requirements. The number of gates between latches used in earlier designs would not work in this methodology. For example, in Gronowski et al. (1998), two possible scenarios are depicted, as shown in Fig. 9.28, which illustrates the capability of buffering, on the left, and conditioning, on the right, of the main clock, subject to a certain set of constraints. Timing analysis first identifies the clock that is common to both the driving and receiving path, shown as D and R in Fig. 9.28. Critical path analysis verifies that the difference in delay between the drive path D and receive path R, including the skew and setup time, does not exceed the phase or cycle times of the common clock. The worst-case analysis takes into account effects that minimize R and maximize D. The races pose a dual problem where the effects that maximize R and minimize D are concerned. For case that path R, including the hold time and skew, is longer than D, a race-through occurs. A detailed analysis of the clocking style and verification methodology is given in Bailey and Benschneider (1998).

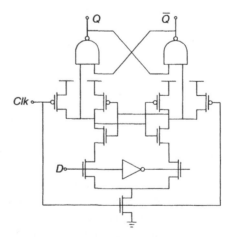

Figure 9.27. The 21264 flip-flop. (Gronowski et al. 1998), Copyright © 1998 IEEE.

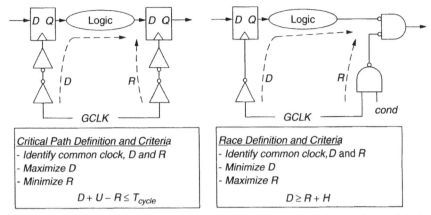

Figure 9.28. Critical-path and race analysis for clock buffering and conditioning. (Gronowski et al. 1998), Copyright © 1998 IEEE.

The fourth generation of Alpha processor did not change the flip-flop methodology, but rather has dealt with the increased area and complexity through the use of the synchronized main clock domains.

Throughout the past eight years and four generations, the Alpha microprocessor continued to deliver the peak performance and leading the industry with new ideas in all aspects of microprocessor design. Of these, the clocking and latch design methodologies were addressed in this section.

One of the interesting conclusions in Bailey and Benschneider (1998) is that in the design of high-performance processors, more and more attention has been paid to the accurate modeling of delay-path variations, clock skew, process, and so forth, in order to be able to predict, nonconservatively, but accurately the behavior of the system, and in that way reliably decrease the operating margin as much as possible. Beside the gains in performance obtained from process scaling, this type of increased level of detail in analysis continually enabled new material for trade-offs and encouraged the creativity of architects and circuit designers.

It is hard to predict what type of clocking and CSE design will be used in future machines. The requirement for speed effectively mandates latchless pipelines with only a few gates per stage. On the other hand, the power-dissipation requirement mandates the use of conditional clocking, which requires reliable static latch operation. The increase in design size dictates clock hierarchy and globally synchronized separate clock domains. In future, even more attention will be paid to the separation of critical paths from the rest of the data path and control and possible application of different clocking methodologies to different sections based on the power-performance requirements.

9.4. CLOCKED STORAGE ELEMENTS IN IBM PROCESSORS

Traditionally all clocked storage elements in IBM processors were required to adhere to LSSD methodology (Williams and Eichelberger 1977). LSSD implicitly prohibits the use of flip-flops. Therefore all the clocked storage elements are level-sensitive latches. Level sensitivity in IBM terminology means that the only mechanism responsible for capturing data is the logic value of the clock signal (level), and not the rate of the clock signal transition. Flip-flop structure is sensitive to variations in the clock rise and fall times, which can present a reliability problem, as discussed in earlier chapters of this book. This fact was recognized at IBM very early, thus resulting in LSSD restrictions on the use of flip-flops. Thus the clocked storage element used in IBM is a "polarity-hold, level-sensitive latch." A logic implementation of a hazard-free polarity-hold latch is shown in Fig. 9.29.

In IBM terminology polarity hold means the ability to maintain the logic level (polarity) of a signal by the clocked storage element (latch, in this case). In the example shown in Fig. 9.29, the value of the data signal is reflected in its true form at the output.

Before going further, it is important to provide a basic explanation of IBM LSSD methodology.

9.4.1. Level-Sensitive Scan Design

The issue of testability is closely related to the latch design and choice of a clocked storage element used in a system. Therefore, LSSD is a design methodology. It was developed at the IBM Corporation and used systematically in all IBM designs (Eichelberger and Williams 1977; Williams and Eichelberger 1977). The origins of LSSD can be traced to the IBM System 360 models and the NEC 2200 model 700, although LSSD was fully implemented for the first time on IBM System 38 (Stolte and Berglund 1979). The origins of scan-based design go even further back in time to the research conducted at Stanford University (Williams and Angel 1973).

LSSD is one solution to the problem of test and test generation for digital systems. The basic idea of LSSD is to convert a sequential network into a combinational network by logically cutting the feedback loops. This logical dissection is performed by converting all storage elements in the Huffman sequential-network model (Fig. 1.3.) into shift register latches and connecting them into one or more shift registers, as shown in Fig. 9.30. At this point it is possible to place the logic network into any desired state by shifting-in the proper values into the

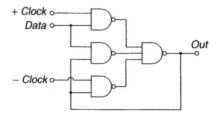

Figure 9.29. Hazard-free level-sensitive polarity-hold latch. (Eichelberger 1983)

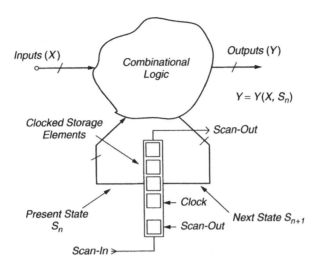

Figure 9.30. General LSSD configuration.

Shift-Register Latches (SRL). It is also possible to scan out any response. Thus, for testing purposes, the network looks like a combinational network, which greatly facilitates test generation.

There are two aspects of LSSD methodology that impact timing and clock. The first attribute is the requirement that the system is *level sensitive*, and the second one is the requirement for a *scan design*.

Level sensitivity is defined in the requirements for the latch design. The latches used are assumed to be reacting to logic voltage levels and not to be affected by the clock transition time. This is consistent with the definition of a latch in this book, as opposed to a flip-flop. Further, clocks are recommended to be nonoverlapping during system operation and are never overlapping during testing. Hence the network is immune to fast paths.

The requirement for scan design is implicit in the requirement that the latches used consist of SRLs, which are interconnected with one or more shift-register chains. Thus, the key capability of scan design is the capability of complete control and to observe all latches used in the system.

These two features are essential in making a sequential network appear like a combinational network. LSSD makes it possible to scan-in, as well as scan-out, values into and from all the latches in the system.

The advantages of LSSD are summarized as follows:

1. System performance is independent of the time-dependent characteristics of the signals, such as rise and fall time.

2. As far as test generation is concerned, all the logic networks are treated as combinational, thus greatly simplifying testing and the test generation process.

3. The ability to scan simplifies the debugging of designs.

4. The ability to scan simplifies the machine bring-up and diagnostic process.

5. Design verification is simplified.

6. In the case where complete systems are designed using LSSD, the same manufacturing tests can be applied to the diagnosis of faults on the customer's site.

There are two basic ways to design logic in LSSD. One is by using a single latch, the other is by using the double-latch design (as described in this chapter). Double-latch design is also known as M–S or latch-trigger design. IBM LSSD SRL is shown in Fig. 9.31.

A Shift-Register Latch is defined as a combination of two latches: a data input latch L_1 and a second latch L_2, which is used in normal, or shift register, operation. Latch L_1 can be fed by one or more system clocks, data inputs, set inputs, reset inputs, scan data inputs and shift-A clock inputs. Latch L_2 only can be fed by latch L_1 and shift-B clock inputs. System data outputs can be taken from latch L_1, from latch L_2, or from both L_1 and L_2. At least one output from L_2 must be used to provide a shift-register data path.

In double-latch design, shown in Fig. 9.32, outputs are taken from the L_2 latches. Since the L_1 and L_2 latches must have separate clocks, this design is inherently level sensitive. During the normal operation the L_1–L_2 (M–S) latch is clocked with the $-C$ and $+B$ clocks. Clock $-C$ is responsible for latching data input into the master latch L_1. In the scan mode L_1–L_2 latches are clocked by $+A$ and $+B$ clocks, and the master latch is latching data from the Scan_In input. All the latches are interconnected into a long chain forming a shift register. The content of this register is scanned out into the tester, and alternatively a new test vector is scanned in.

Double-latch design requires no more than two system clocks, C_1 and C_2, and two shift clocks, A and B. The C_2 clock for the L_2 latch behaves like a shift B clock during testing and a system clock C_2 during normal operation. It is not necessary to use two separate clocks, C_2 and B, since the function can be shared during the normal operation and testing.

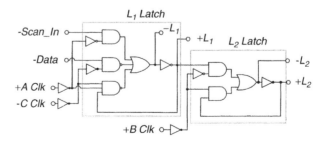

Figure 9.31. LSSD shift register latch.

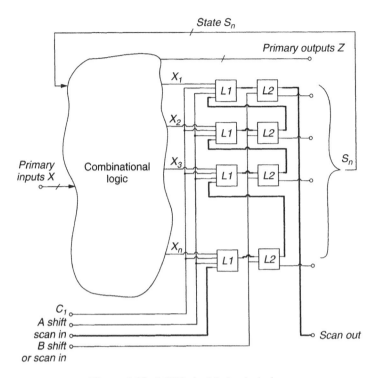

Figure 9.32. LSSD double-latch design.

LSSD is a concept that can be applied to a complete system design from the module or a card to a chip.

9.4.2. Examples of Clocked Storage Elements

IBM S/390 G4 Parallel Server Processor   There are two types of latches used in the IBM S/390 G4 processor (Sigal et al. 1997): a single L_2 latch and a L_1–L_2 pair. There are no midcycle latches (split latches) used, in spite of L_2 being a single latch. For each type of latch, there is a corresponding clock block whose purpose is to generate local clocks for the latches. The first combination of the latch and the local clock generator is shown in Fig. 9.33. A short local clock pulse *CLKL* is generated from the global clock *CLKG* following the trailing edge of the global clock. To create *CLKL*, the principle of *reconvergent fan-out* with *nonequal parities of inversion* (five in this case) is used on the *CKLG*. This generates a short negative pulse of approximately six inverter delays, which is used as a local clock. The local clock, *CLKL*, is clocking a domino style multiplexer.

During the normal operation, the local clock is enabled and it is used to clock the first stage (master latch), consisting of a domino-style multiplexer. Various inputs could be latched into the master latch, depending on the state

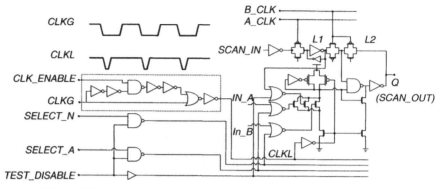

Figure 9.33. LSSD SRL with multiplexer used in the IBM S/390 G4 processor. (Sigal et al. 1997), reproduced by permission.

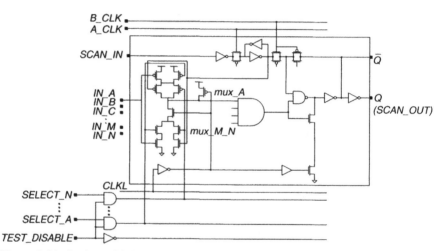

Figure 9.34. Static multiplexer version of the SRL used in the IBM S/390 G4. (Sigal et al. 1997), reproduced by permission.

of the processor's control signals. The second stage latch is a slave latch, thus the M–S pair consists of an input domino multiplexer and an L_2 latch. For this operation to go undisturbed, both the A_CLK and B_CLK signals are held at logic-0 level. In the test mode, the system latches are connected into a scan chain. The $TEST\_DISABLE$ signal is held at the logic-0 level, thus enabling the value from the $SCAN\_IN$ input. Scaning the test vectors in and out of the system is accomplished by asserting the A_CLK and B_CLK signals. In addition, $CLKL$ transfers the value from the L_1 scan latch into the domino master latch. Both dynamic and static implementation of the input multiplexer are attainable. A static multiplexer version of the multiplexer, SRL, used in the IBM S/390 G4 is shown in Fig. 9.34.

The second clocked storage element is used in non-timing-critical data-flow macros and in control macros where all latches are single-input and the speed advantage of an L_2-only latch is reduced. The local clock block generates C_1/C_2 clocks. The clock overlap between C_1 and C_2 is kept close to 0. However, it is possible to create a positive overlap between the C_1 and C_2 clocks in order to increase system performance by reducing the latch propagation delay. This requires padding the fast signals, as discussed in the previous chapters. The second clocked storage element is shown in Fig. 9.35. It consists of a relatively simple M–S L_1/L_2 latch combination and a local clock generator responsible for generating C_1 and C_2, which are two-phase nonoverlapping clocks.

In order to have better control of clocks in the S/390 G4 processor, several clock-generating elements were used. Their purpose is to provide different phasing of the C_1/C_2 clocks. The clock generator shown in Fig. 9.36. is used to provide separation (nonoverlap) between the C_1 and C_2 clocks in order to

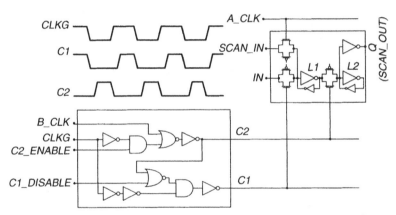

Figure 9.35. A clocked storage element is used in the non-timing-critical timing macros of the IBM S/390 G4 processor. (Sigal et al. 1997), reproduced by permission.

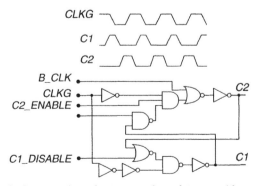

Figure 9.36. The clock-generation element used to detect problems created with fast paths: IBM S/390 G4 processor. (Sigal et al. 1997), reproduced by permission.

detect problems created with fast paths. Alternatively, it is possible to delay the C_1 falling edge from its nominal value to examine how much margin exists for fast paths. Another circuit delays both the C_1 and C_2 clocks from their nominal values, thus allowing for cycle stealing (time borrowing) from the previous cycle.

IBM PowerPC   The IBM experimental processor, which was the first one to reach the 1-GHz mark (Silberman et al. 1998), uses multiplexed input latches in order to merge the important logic operation with the storage function (Fig. 9.37b). The latches provide data-input ports, hold-input, and a scan-input port for full scan testing. The inputs take single-rail static or a dynamic signal and generate dual-rail pulsed outputs for driving dynamic logic. The L_1 latch is a differential structure driving the L_2 latch, which is also used for scan output. The scan-select input has priority over other mux-select inputs, as shown in the Fig. 9.37a.

IBM PowerPC 603   The IBM PowerPC 603™, which was designed under a cooperation agreement between IBM, Motorola, and Apple Computer, uses another standard IBM approach to clocking and design methodology under compliance with LSSD. It represents a classic M–S $(L_1–L_2)$ structure clocked by two separate clocks, C_1 and C_2, and the *ACLK* clock, which is used during the scan mode. A schematic diagram of PowerPC 603 M–S latch (Gerosa et al.

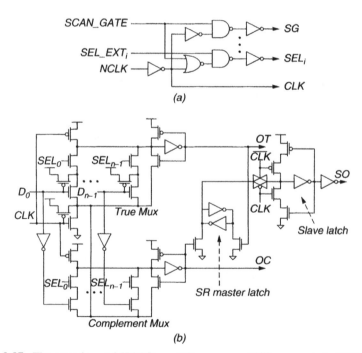

Figure 9.37. The experimental IBM PowerPC processor. (Silberman et al. 1998), reproduced by permission.

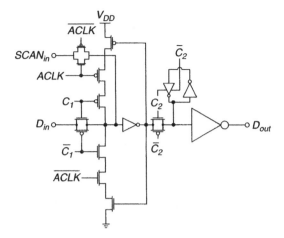

Figure 9.38. The PowerPC 603 MSL. (Gerosa et al. 1994), Copyright © 1994 IEEE.

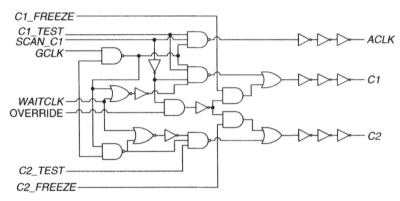

Figure 9.39. The PowerPC 603 local clock regenerator. (Gerosa et al. 1994), Copyright © 1994 IEEE.

1994) is shown in Fig. 9.38. The characteristics of this clocked storage element have been examined in earlier chapters.

A more interesting aspect is a local clock regenerator, which is used to generate the local C_1 and C_2 clocks from a global clock signal clock, shown in Fig. 9.39. This clock regenerator provides electrically correct local clock signals, as well as test clocks and processor power management control features. The outputs of the local clock regenerator are master and slave latch clock signals, C_1 and C_2, respectively, and the scan port clock *ACLK*. The input to the local clock regenerator is global clock signal, *GCLK*, which is the main clock signal distributed across the PowerPC 603 chip. Shutting off the local clocks is possible using the *OVERRIDE* signal. This is used for static power management in order to reduce power. Test control is accomplished by injecting the C_1_TEST,

C_2_TEST, and $SCAN\_C_1$ signals. For local power management, local clock signals can be frozen by using C_1_FREEZE and C_2_FREEZE controls. $WAITCLK$ input has the function of providing additional separation between the C_1 and C_2 clock signals, in the event of unanticipated race conditions.

IBM Power4 Microprocessor   The IBM Power4™ provides processing power for the IBM eServer p690, which is an IBM high-end 64-bit POWER™ architecture. The server can be configured as an 8-to-32-way server system. The microprocessor is implemented using 174-million transistors, and it runs at a frequency higher than 1.3 GHz. The processor, shown in Fig. 9.40, contains two microprocessor cores, high-speed busses, and an on-chip memory subsystem. It is fabricated in state-of-the-art IBM 0.18 μ CMOS silicon-on-insulator (SOI) technology, with seven levels of copper wiring. The IBM Power4 processor uses novel clocking and latches, which were necessary in order to achieve such a high-frequency of operation (Warnock et al. 2002).

A high-quality global clock signal distributed to every latch and clocked circuit was essential. The global clock distribution is especially challenging for a large and complex chip because of the longer wires and the gain needed to drive the large distributed clock load.

Latch Design   In keeping with IBM tradition, the majority of the clocked storage elements used are traditional MSLs. The scan input was brought into the keeper

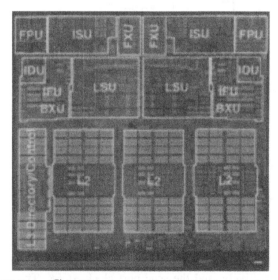

Figure 9.40. IBM Power4™ 64-bit processor used in IBM eServer p690. The microprocessor consists of 174 million transistors and runs at 1.3 GHz, contains two microprocessor cores, and an on-chip memory subsystem. It is fabricated in state-of-the-art IBM 0.18 μ CMOS SOI technology with seven levels of copper wiring (Warnock et al. 2002), reproduced by permission.

latch, thus minimizing its impact on latch delay. The MSLs were designed to be able to tolerate a certain amount of clock uncertainty. In order to minimize the pipeline overhead imposed by the latch, designers were allowed to customize the logic gate, which drives the master latch, as shown in Fig. 9.41. All the latches were sized in order to separately control and optimize the latch power, setup time, and clock-to-data-out delay.

The two local clock phases (c_1 and c_2), as well as the scan clock, were derived locally from one tap of the global clock, in the way shown in Fig. 9.42.

Each local clock signal generator and buffer is controlled by two control inputs for test and debug capability: "c_1_Stop" and "Scanclk_Stop". For protection against race conditions, two other control signals: "Local_u" and "Global_u" were provided. The "Local_u" signal is used to delay the rising edge of the clock, and is controlled by the designer. The "Global_u" signals were used to selectively delay the rising edge of the clock for debugging purposes, and are controlled by scan latches. The c_1 and scan clock buffers had separate stop controls, allowing arbitrary sequencing of the scan and c_1 (system) clocks, while the c_2 clock was free-running, with the stop signal tied to ground.

In order to reduce the overall overhead of the CSE and absorb the clock skew and process parameters variability in the across-chip line-width variation, "split-latch" design, shown in Fig. 9.43, was allowed. This design style already has been discussed in Section 4.2, and an example was provided in Section 4.3 (Fig. 4.2 and Fig. 4.5). This design style allows the logic signals on critical paths to propagate through alternating cycle-boundary (master, or c_1) and mid-cycle (slave, or c_2) latches without incurring a setup time penalty. On average, a half-cycle of logic is allowed between the c_1 and c_2 latches or between the c_2 and c_1 latches. Less logic between any two latches means that time is given up to the logic following the receiving latch, and more logic means that time is taken from the following logic (Warnock et al. 2002).

However, the area overhead for LSSD compatibility becomes significant in this case, since an additional c_2 latch must be provided (aside from the separate c_2

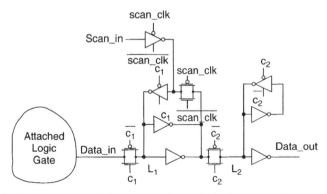

Figure 9.41. Standard transmission-gate MSL with LSSD capability. (Warnock et al. 2002), reproduced by permission.

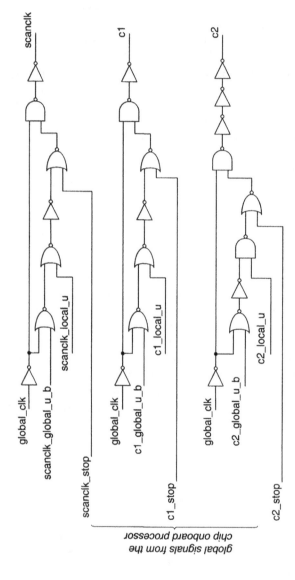

Figure 9.42. Standard transmission-gate MSL with LSSD capability. (Warnock et al. 2002), reproduced by permission.

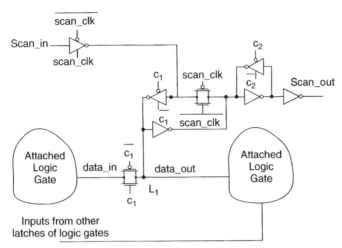

Figure 9.43. Scannable split latch with LSSD capability used in the IBM Power4™. Designers were allowed to tune transmission gate size and specify input and output gates (Warnock et al. 2002), reproduced by permission.

added to the downstream logic) for scan functionality. Even in this situation, the extra area was a relatively small addition to the overall total, and the flexibility of this scheme would often allow area savings in other parts of the design (Anderson et al. 2001).

Aside from the benefits offered by split-latch design, the IBM Power4 team found some drawbacks in this design as well, such as the increased difficulty of timing paths through logic containing these latches. The timing tool had to be able to deal with multicycle paths through transparent latches, including loops and other difficult topological situations, and then had to present the timing data in an intelligible way. In addition, there were issues with testing at speed, including the fact that it became difficult to assess how many back-to-back cycles would be needed to capture all of the critical timing paths through the machine. Also, timing failures could become much more difficult to debug.

A number of special latch cells with an integrated logic were built into the design library. These cells allowed the logic to be merged with the latch cell, and avoided exposing the latch transmission-gate input to potentially noisy wires. These cells are shown in Fig. 9.44.

Split-latch designs were available with an integrated front-and-back logic gate, as shown in Fig. 9.45.

These solutions allowed the designer to have almost the same resources as the custom designer in order to minimize the latch overhead and provide for a clock-skew-tolerant operation.

In summary, IBM microprocessors use a conservative design with a strong emphasis on testability and reliability, which has been IBM's trademark over the years. All the latches used in IBM products are required to be LSSD compatible,

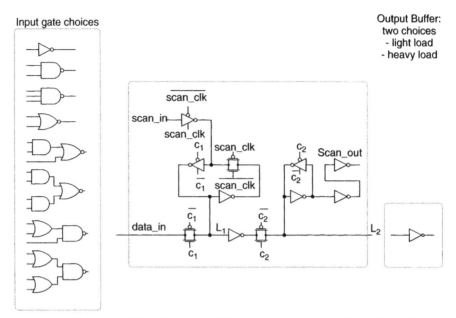

Figure 9.44. Library MSL with integrated front-end logic gate choices. (Warnock et al. 2002), reproduced by permission.

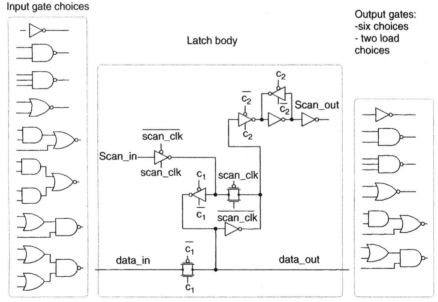

Figure 9.45. Split-latch designs with integrated front- and back-end logic-gate choices. (Warnock et al. 2002), reproduced by permission.

thus to incorporate scan. The diagnostic and machine bring-up phases are given equal importance at IBM, as shown by various clock-edge and clock-overlap control signals that were extensively used. However, in spite of the emphasis on testability, reliability, and availability, IBM designs are capable of achieving remarkable speed, thus placing IBM microprocessors in the performance lead.

REFERENCES

Afghahi, M., and Svensson, C. (1990). "A Unified Single-Phase Clocking Scheme for VLSI Systems." *IEEE Journal of Solid-State Circuits*, vol. SC-25 (no. 1), February, p. 225–33.

Amdahl, G. M. (1964). "The Structure of System /360 Part III: Processing Unit Design Considerations." *IBM Systems Journal*, vol. 3 (no. 2), p. 144–64.

Anderson, C. J., Petrovick, J., Keaty, J. M., Warnock, J., Nusbaum, G., Tendler, J. M., Carter, C., Chu, S., Clabes, J., DiLullo, J., Dudley, P., Harvey, P., Krauter, B., LeBlanc, J., Lu, P.-F., McCredie, B., Plum, G., Restle, P., Runyon, S., Scheuermann, M., Schmidt, S., Wagoner, J., Weiss, R., Weitzel, S., and Zoric, B. (2001). "Physical Design of a Fourth-Generation POWER GHz Microprocessor," *IEEE International Solid-State Circuits Conference, Digest of Technical Papers*, p. 232–3.

Anderson, D. W., Sparacio, F. J., and Tomasulo, R. M. (1967). "The IBM System/360 Model 91: Machine Philosophy and Instruction Handling," *IBM Journal of Research and Development*, vol. 11 (no. 1), p. 8–24.

Baeg, S., and Rogers, W. A. (1999). "A Cost-Effective Design for Testability: Clock Line Control and Test Generation Using Selective Clocking," *IEEE Transactions on Computer-Aided Design of Integrated Circuits and Systems*, vol. 18 (no. 6), June, p. 850–61.

Bailey, D. W., and Benschneider, B. J. (1998). "Clocking Design and Analysis for a 600-MHz Alpha Microprocessor," *IEEE Journal of Solid-State Circuits*, vol. SC-33 (no. 11), November.

Benschneider, B. J. et al. (1995). "A 300-MHz 64-b Quad-Issue CMOS RISC Microprocessor," *IEEE Journal of Solid-State Circuits*, vol. SC-30 (no. 11), November.

Bhagwan, R., and Rogers, A. (1997). "A 1-GHz Dual-Loop Microprocessor PLL with Instant Frequency Shifting." *ISSCC Digest of Technical Papers*, p. 336–7.

Bowhill, W. J. et al. (1995). "A 300 MHz 64b Quad-Issue CMOS RISC Micro-processor," *IEEE International Solid-State Circuits Conference*, vol. XXXVIII, February, p. 182–3.

Burd, T., Pering, T., and Brodersen, R. W. (2000). "A Dynamic Voltage Scaled Microprocessor System." *Proceedings of ISSCC 2000*, p. 294–5.

Chandrakasan, A. (1994). *Low-Power Digital CMOS Design*, Ph.D. Thesis, University of California, Berkeley.

Chandrakasan, A., Bowhill, B. J., and Fox, F. (2001). *Design of High-Performance Microprocessor Circuits*. Piscataway, NJ: IEEE Press.

Charnas, A. et al. (1995). A 64b Microprocessor with Multimedia Support, *ISSCC Digest of Technical Papers*, p. 178–9.

Cray Research. (1984). *S Series Mainframe Reference Manual HR 0029*. Minneapolis, MN: Cray Research.

Dobberpuhl, D. W. (1997). "Circuits and Technology for Digital's StrongARM and ALPHA Microprocessors." *Proceedings of the Seventeenth Conference on Advanced Research in VLSI* (Ann Arbor, Michigan, September 15–16), p. 2–11.

Dobberpuhl, D. W. et al. (1992). "A 200-MHz 64-b Dual-Issue CMOS Micro-processor," *IEEE Journal of Solid-State Circuits*, vol. SC-27, November, p. 1555–67.

Earl, J. (1965). "Latched Cary-Save Adder," *IBM Technical Disclosure Bulletin*, vol. 7 (no. 10), March, p. 909–10.

Eichelberger, E. B. (1983). Latch Design Using Level Sensitive Scan Design, Proceedings of COMPCON, San Francisco.

Eichelberger, E. B., and Williams, T. W. (1977). "A Logic Design Structure for LSI Testing," *Proceedings of the 14th Design Automation Conference*.

Fishburn, J. P. (1990). "Clock Skew Optimization," *IEEE Transactions on Computers*, vol. 39, p. 945–51.

Flynn, M. J. (1966). "Very High-Speed Computing System," *Proceedings of the IEEE*, vol. 54 (no. 12), December, p. 1901–09.

Flynn, M. J., and Amdahl, G. M. (1965). "Engineering Aspects of Large High Speed Computer Design," *Proceedings of the Symposium on Microelectonics and Large Systems*, Washington, DC: Spartan Press.

Friedman, E. (1995). *Clock Distribution Networks in VLSI Circuits and Systems*. New York: IEEE Press.

Furber, S. B., Efthymiou, A., Garside, J. D., Lloyd, D. W., Lewis, M. J. G., and Temple, S. (2001). "Power Management in the Amulet Microprocessors," *IEEE Design & Test of Computers*, vol. 18 (no. 2), March/April.

Gago, A., Escano, R., and Hidalgo, J. A. (1993). "Reduced implementation of D-type DET Flip-Flops," *IEEE Journal of Solid-State Circuits*, vol. SC-28 (no. 3), March p. 400–2.

Gardner, F. (1979). *Phase Lock Techniques*. New York: Wiley.

Geannopoulos, G., and Dai, X. (1998) "An Adaptive Digital Deskewing Circuit for Clock Distribution Networks," *ISSCC Digest of Techn. Papers*, pp. 400–1.

Gerosa, G., Gary, S., Dietz, C., Dac, P., Hoover, K., Alvarez, J., et al. (1994). "A 2.2 W, 80 MHz Superscalar RISC Microprocessor," *IEEE Journal of Solid-State Circuits*, vol. SC-29, (no. 12), December.

Gieseke, B. A. et al. (1991) "Push-Pull Cascode Logic," U.S. Patent No. 5,023,480 (June).

Gieseke, B. A. et al. (1997). "A 600MHz Superscalar RISC Microprocessor with Out-of-Order Execution," *Proceedings of the IEEE International Solid-State Circuits Conference*, vol. XL, February, p. 176–7.

Goncalves, N. F., and DeMan, H. J. (1983). "NORA: A Racefree Dynamic CMOS Technique for Pipelined Logic Structures, *IEEE Journal of Solid-State Circuits*, vol. SC-18 (no. 3), June.

Gowan, M. K., Biro, L. L., and Jackson, D. B. (1998). "Power Considerations in the Design of the Alpha 21264 Microprocessor," in *Proc. DAC 1998*, p. 726–31.

Greenhill, D. et al. (1997). "A 330MHz 4-Way Superscalar Microprocessor." *ISSCC Digest of Technical Papers*, p. 166–7.

Gronowski, P. E., Bowhill, W. J., Preston, R. P., Gowan, M. K., and Allmon, R. L. (1998). "High-performance Microprocessor Design," *IEEE Journal of Solid-State Circuits*, vol. 33, May, p. 676–86.

Hajimiri, A. (1998). Jitter and Phase Noise in Electrical Oscillators, Ph.D. Thesis, Stanford University, Stanford, CA.

Halin, T. G., and Flynn, M. J. (1972). "Pipelining of Arithmetic Functions," *IEEE Transactions on Computers*, vol. C-21 (no. 8), August, p. 880–6.

Hamada, M., Terazawa, T., Higashi, T., Kitabayashi, S., Mita, S., Watanabe, Y., Ashino, M., Hara, H., and Kuroda, T. (1999). "Flip-Flop Selection Technique for Power-Delay Trade-Off," *Proceedings of the IEEE International Solid-State Circuits Conference*, vol. XLII, February, p. 270–1.

Harris, D., and Horowitz, M. (1997). "Skew-Tolerant Domino Circuits," *IEEE Journal of Solid-State Circuits*, vol. SC-32, p. 1702–11.

Harris, D., Huang, S. C., Nadir, J., Chu, C.-H., Stinson, J. C., and Ilkbahar, A. (1996). "Opportunistic Time-Borrowing Domino Logic," U.S. Patent No. 5,517,136.

Hashimoto, M., Onodera, H., and Tamaru, K. (1998) "A Power Optimization Method Considering Glitch Reduction by Gate Sizing," in *ISPLED Digest of Technical Papers* (August), p. 221–6.

Heald, R. et al. (2000a). "A Third Generation SPARC V9 Microprocessor," *IEEE Journal of Solid-State Circuits*, vol. 35, p. 1526–38.

Heald, R. et al. (2000b). "Implementation of a 3rd Generation SPARC V9 64b Microprocessor," *ISSCC Digest of Technical Papers*, p. 412–3.

Heo, S., and Asanovic, K. "Load-Sensitive Flip-Flop Characterization," *Proc. IEEE Workshop on VLSI*, (Orlando, Fl, April), p. 87–92.

Hofstee, P. et al. (2000). "A 1-GHz Single-Issue 64b PowerPC Processor," in *ISSCC Digest of Technical Papers*, p. 92–3.

Jain, A. et al. (2001). "A 1.2 GHz Alpha Microprocessor with 44.8 GB/s Chip Pin Bandwidth," *Proceedings of the IEEE International Solid-State Circuits Conference*, February, p. 240–1.

Kawaguchi, H., and Sakurai, T. (1998). "A Reduced Clock-Swing Flip-Flop (RCSFF) for 63% Power Reduction," *IEEE J Journal of Solid-State Circuits*, vol. SC-33, (no. 5), May, p. 807–11.

Kim, B., Weigandt, T., and Gray, P. (1994). "PLL/DLL System Noise Analysis for Low Jitter Clock Synthesizer Design," *Proceedings of the 1994 International Symposium on Circuits and Systems*, vol. 4, p. 31–8.

Kitahara, T. et al., (1998). "A Clock-Gating Method for Low-Power LSI Design," *Proceedings of the ASC-DAC'98, Conference*, p. 307–12.

Klass, F. (1998). "Semi-Dynamic and Dynamic Flip-Flops with Embedded Logic," *Symposium on VLSI Circuits, Digest of Technical Papers* (June), p. 108–9.

Klass, F. et al. (1999). "A New Family of Semidynamic and Dynamic Flip-Flops with Embedded Logic for High-Performance Processors," *IEEE Journal of Solid-State Circuits*, vol. SC-34, (no. 5), May, p. 712–16.

Kogge, P. (1981). *The Architecture of Pipelined Computers*. New York: McGraw-Hill.

Kojima, H., Tanaka, S., and Sasaki, K. (1995). "Half-Swing Clocking Scheme for 75% Power Savings in Clocking Circuitry," *IEEE Journal of Solid-State Circuits*, vol. SC-30, (no. 4), April, p. 432–5.

Kong, B.-S., Kim, S.-S., and Jun, Y.-H. (2000). "Conditional Capture Flip-Flop for Statistical Power Reduction," in *ISSCC Digest of Technical Papers*, p. 290–1.

Kozu, S. et al. (1996). "A 100 MHz, 0.4 W RISC Processor with 200 MHz Multiply Adder, Using Pulse-Register Technique," *Digest of Technical Papers, 1996 IEEE International Solid-State Circuits Conference* (San Francisco, February 8–10), p. 40–1.

Kurd, N. A. et al. (2001). "Multi-GHz Clocking Scheme for Inel®Pentium® 4 Microprocessor," *ISSCC Digest of Technical Papers*, p. 404–5.

Lauterbach, G. et al. (2000). "UltraSPARC-III: A 3rd Generation 64b SPARC Microprocessor," *ISSCC Digest of Technical Papers*, p. 410–11.

Lev, L. A. et al. (1995). "A 64Mb Microprocessor with Multimedia Support," *IEEE Journal of Solid-State Circuits*, vol. SC-30, p. 1727–38.

Lim, K., Park, C., Kim, D., and Kim, B. (2000). "A Low-Noise Phase-Locked Loop Design by Loop Bandwidth Optimization," *IEEE Journal of Solid-State Circuits*, vol. SC-35, June, p. 807–15.

Lin, I., Ludwig, I. A., and Eng, K. (1992). "Analyzing and Cycle Stealing on Synchronous Circuits with Level-Sensitive Latches" *Proceedings of the ACM/IEEE Design Automation Conference*, p. 393–8.

Llopis, R. P., and Sachdev, M. (1996). "Low Power, Testable Dual Edge Triggered Flip-Flops," *Proceedings of the International Symposium on Low Power Electronics and Design*, p. 341–5.

LSSD Rules and Applications. (1985). Manual 3531, Release 59.0, IBM Corporation, March 29.

Madden, W. C., and Bowhill, W. J. (1990). "High Input Impedance, Strobed Sense-Amplifier," U.S. Patent No. 110. 4,910,713, (March).

Mansuri, M., Yang, C.-K. K. (2002). "Jitter Optimization Based on Phase-Locked Loop Design Parameters," *IEEE International Solid-State Circuits Conference*, vol. 1, February, p. 138–9.

Markovic, D., Nikolic, B., and Brodersen, R. W. (2001). "Analysis and Design of Low-Energy Flip-Flops," in *ISPLED Digest of Technical Papers*, August, p. 52–5.

Matsui, M., Hara, H., Uetani, Y., Kim, L., Nagamatsu, T., Watanabe, Y., Chiba, A., Matsuda, K., and Sakurai, T. (1994). "A 200 MHz 13 mm^{2}2-D DCT Macrocell Using Sense-Amplifying Pipeline Flip-Flop Scheme," *IEEE Journal of Solid-State Circuits*, vol. SC-29, December p. 1482–90.

Montanaro, J., Witek, R. T., Anne, K., Black, A. J., Cooper, E. M., Dobberpuhl, D., Donahue, P. M., Eno, J., Farell, A., Hoeppner, G., Kruckemyer, D., Lee, T. H., Lin, P., Madden, L., Murray, D., Pearce, M., Santhanam, S., Snyder, K. J., Stephany, R., and Thierauf, S. C., (1996). "A 160MHz 32b 0.5W CMOS RISC Microprocessor," *Proceedings of the IEEE International Solid-State Circuits Conference*, vol. XXXIX, February, p. 214–15.

Montanaro, J., Witek, R. T., Anne, K., Black, A. J., Cooper, E. M., Dobberpuhl, D. W., Donahue, P. M., Eno, J., Hoeppner, G. W., Kruckemyer, D., Lee, T. H., Lin, P. C. M., Madden, L., Murray, D., Pearce, M. H., Santhanam, S., Snyder, K. J., Stephany, R., and Thierauf, S. C. (1997). "A 160-MHz, 32-b, 0.5-W CMOS RISC Microprocessor," *Digital Technical Journal*, vol. 9, (no. 1), Digital Equipment Corp. p. 49–62.

Nikolic, B., and Oklobdzija, V. G. (1999). "Design and Optimization of Sense Amplifier-Based Flip-Flops," *Proceedings of the 25th European Solid-State Circuits Conference, ESSCIRC'99*, (Duisburg, Germany, September 21–23), p. 410–13.

Nedovic, N., and Oklobdzija, V. G. (2000a). "Dynamic Flip-Flop with Improved Power," *Proceedings of the IEEE International Conference on Computer Design: VLSI in Computers and Processors* (Austin, TX, September 17–20), p. 323–6.

Nedovic, N., and Oklobdzija, V. G. (2000b). "Hybrid Latch Flip-Flop with Improved Power Efficiency," *Proceedings of the 13th Symposium on Integrated Circuits and Systems Design*, (Manaus, Brazil, September 18–24), p. 211–15.

Nedovic, N., Aleksic, M., and Oklobdzija, V. G. (2001). "Timing Characterization of Dual-Edge Triggered Flip-Flops," *Proceedings of the International Conference on Computer Design, ICCD 2001*, Austin, TX, (September 23–26).

Nedovic, N., Oklobdzija, V. G., Walker, W. W., and Aleksic, M. (2002). "A Low Power Symmetrically Pulsed Dual Edge-Triggered Flip-Flop," *IEEE European Solid-State Circuits Conference*, September.

Nikolic, B., Stojanovic, V., Oklobdzija, V. G., Jia, W., and Leung, M. (1999). "Sense Amplifier-Based Flip-Flop," in *ISSCC Digest of Technical Papers*, p. 282–3.

Nogawa, M., and Ohtomo, Y. "A Data-Transition Look-Ahead DFF Circuit for Statistical Reduction in Power Consumption," *IEEE Journal of Solid-State Circuits*, vol. SC-33, (May), p. 702–6.

Nogawa, M., and Ohtomo, Y. (1998). "A Data-Transition Look-Ahead DFF Circuit for Statistical Reduction in Power Consumption," *IEEE Journal of Solid-State Circuits*, vol. SC-33, (no. 5), May, p. 702–6.

Oklobdzija, V. G. (1999). *High-Performance System Design: Circuits and Logic*, New York: IEEE Press.

Oklobdzija, V. G., Stojanovic, V. (2001). "FLIP-FLOP," U.S. Patent No. 6,232,810 (May 15).

Partovi, H. et al. (1996). "Flow-Through Latch and Edge-Triggered Flip-Flop Hybrid Elements," *1996 IEEE International Solid-State Circuits Conference. Digest of Technical Papers*, (San Francisco, February 8–10).

Razavi, B., Ed. (1996). *Monolithic Phase-Locked Loops and Clock Recovery Circuits — Theory and Design*. New York: IEEE Press.

Rusu, S., and Tam, S. (2000). "Clock Generation and Distribution for the First IA-64 Microprocessor," *Proceedings of the IEEE International Solid-State Circuits Conference*, vol. XLIII, February, p. 176–7.

Saint-Laurent, M. et al. (2002). "Optimal Sequencing Energy Allocation for CMOS Integrated Systems," *Proceedings of the International Symposium on Quality Electronic Design*, p. 94–9.

Sakallah, K. A., Mudge, T. N., and Olukotun, O. A. (1992). "Analysis and Design of Latch-Controlled Synchronous Digital Circuits," *IEEE Transactions on Computer-Aided Design of Integrated Circuits and Systems*, vol. 11, p. 322–33.

Schutz, J., and Wallace, R. (1998). "A 450MHz IA32 P6 Family Microprocessor," *ISSCC Dig. Tech. Papers*, p. 236–7.

Sidiropoulos, S., and Horowitz, M. A. (1997). "A Semidigital Dual Delay-Locked Loop," *IEEE Journal of Solid-State Circuits*, vol. SC-32, November, p. 1683–92.

Sidiropoulos, S., Liu, D., Kim, J., Wei, G., and Horowitz, M. (2000). "Adaptive Bandwidth DLLs and PLLs Using Regulated Supply CMOS Buffers," *Proceedings of the IEEE Symposium on VLSI Circuits*, June, p. 124–7.

Siewiorek, D. P., Bell, G. C., and Newell, A. (1982). *Computer Structures: Principles and Examples*. McGraw-Hill, New York 1982.

Sigal, L., Warnock, J. D., Curran, B. W., Chan, Y. H., Camporese, P. J., Mayo, M. D., Huott, W. V., Knebel, D. R., Chuang, C. T., Eckhard, J. P., Wu, P. T. (1997). "Circuits Design Techniques for the High Performance CMOS IBM S/390 Parallel Enterprise Server G4 Microprocessor," *IBM Journal of Research and Development*, vol. 41 (no. 4–5), July–September.

Silberman, J., Aoki, N., Boerstler, D., Burns, J. L., Sang, D., Essbaum, A., Ghoshal, U., Heidel, D., Hofstee, P., Kyung, T. L., Meltzer, D., Hung, N., Nowka, K., Posluszny, S., Takahashi, O., Vo, I., Zoric, B. (1998). "A 1.0-GHz Single-Issue 64-bit PowerPC Integer Processor," *IEEE Journal of Solid-State Circuits*, vol. SC-33 (no. 11).

Stojanovic, V., Oklobdzija, V. G. (1999). "Comparative Analysis of Master-Slave Latches and Flip-Flops for High-Performance and Low-Power Systems," *IEEE Journal of Solid-State Circuits,* vol. 34 (no.4), April, p. 536–48.

Stolte, L. A., and Berglund, N. C. (1979). "Design for Testability for the IBM System/38," *Digest of Papers 1979 IEEE Test Conference*, Cherry Hill. NJ.

Strollo, A. G. M., Napoli, E., De Caro, D. (2000). "New Clock-Gating Techniques for Low-Power Flip-Flops," in *ISLPED Digest of Technical Papers*, August, p. 114–9.

Sutherland, I. E., and Sproull, R. F. (1991). "Logical Effort: Designing for Speed on the Back of an Envelope," *Advanced Research in VLSI*, ARVLSI'91, Santa Cruz, CA.

Sutherland, I., Sproull, B., and Harris, D. 1999. *Logical Effort: Designing Fast CMOS Circuits*, Morgan Kaufmann, (Web enhancements from www.mkp.com or ftp://ftp.mkp.com/Logical Effort/CAT Tool).

Suzuki, Y., Odagawa, K., and Abe, T. (1973). "Clocked CMOS Calculator Circuitry," *IEEE Solid-State Circuits*, vol. SC-8, December, p. 462–9.

Svensson, C., and Yuan, J. (1998). "Latches and Flip-Flops for Low Power Systems," in *Low Power CMOS Design*, A. Chandrakasan and R. Brodersen, Eds. Piscataway, NJ: IEEE Press, p. 233–8.

Texas Instruments. (1984). *The TTL Data Book for Design Engineers*, Dallas: Texas Instruments.

Tschanz, J., Siva, N., Zhanping, C., Shekhar, B., Manoj, S., Vivek, D. (2001). "Comparative Delay and Energy of Single Edge-Triggered & Dual

Edge-Triggered Pulsed Flip-Flops for High-Performance Microprocessors," *Proceedings of the 2001 International Symposium on Low Power Electronics and Design* (Huntington Beach, CA, August 6–7).

Unger, S. H., and Tan, C. (1986). "Clocking Schemes for High-Speed Digital Systems," *IEEE Transactions on Computers*, vol. C-35 (no. 10), October.

Von Kaenel, V., Aebischer, D., Van Dongen, R., and Piguet, C. (1998). "A 600 MHz CMOS PLL Microprocessor Clock Generator with a 1.2GHz VCO," *Proceedings of the IEEE International Solid-State Circuits Conference*, vol. XLI, February, p. 396–7.

Wagner, K. (1988). "Clock System Design," *IEEE Design & Test of Computers*, October.

Warnock, J. D. et al. (2007). "The Circuit and Physical Dedign of the POWER4 Microprocessor," *IBM Journal of Research and Development*, vol. 46, (no. 1), January.

Williams, M. J. Y., Angel, J. B. (1973). "Enhancing Testability of Large Scale Integrated Circuits via Test Points and Additional Logic," *IEEE Transactions on Computers*, vol. C-22.

Williams, T. W., Eichelberger, E. B. (1977). "Random Patterns Within a Structured Sequential Logic Design," 1977 and Semiconductor Test Symposium, (Cherry Hill, NY, October 25–27).

Wolf, S. (1995). *Silicon Processing for the VLSI Era, vol. 3, The Submicron MOSFET*, Sunset Beach, CA: Lattice Press.

Woods, J. V., Day, P., Furber, S. B., Garside, J. D., Paver, N. C., and Temple, S. (1997). "AMULET1: An Asynchronous ARM Microprocessor," *IEEE Transactions on Computers*, vol. C-46 (no. 4), April, p. 385–98.

Xanthopoulos, T., Bailey, D. W., Gangwar, A. K., Gowan, M. K., Jain, A. K., and Prewitt, B. K. (2001). "The Design and Analysis of the Clock Distribution Network for a 1.2 GHz Alpha Microprocessor," *Proceedings of the IEEE International Solid-State Circuits Conference*, February, p. 402–3.

Yano, K. et al. (1990) "A 3.8 ns CMOS 16×16-b Multiplier Using Complementary Pass-Transistor Logic," *IEEE Journal of Solid-State Circuits*, vol. SC-25, April, p. 388–385.

Young, I. A., Greason, J. K., and Wong, K.L. (1992). "A PLL Clock Generator with 5 to 10 MHz of Lock Range for Microprocessors," *IEEE Journal of Solid-State Circuits*, vol. SC-27, November, p. 1599–1607.

Young, I. A. et al. (1997). "A 0.35μm CMOS 3-880MHz PLL N/2 Clock Multiplier and Distribution Network with Low Jitter for Microprocessors," ISSCC *Digest of Technical Papers*, p. 330–1.

Yuan, J., and Svensson, C. (1989). "High-Speed CMOS Circuit Technique," *Journal of Solid-State Circuits*, vol. SC-24 (no. 1) February.

Zyuban, V., and Kogge, P. (1999). "Application of STD to Latch-Power Estimation," *IEEE Transactions on Very Large Scale Integration (VLSI) Systems*, vol. 7 (no. I), March, p. 111–5.

INDEX

Active deskewing, 84. *See also* Deskewing
Alpha, *see* Microprocessor
Alpha particles, 204
Asynchronous systems, 4

C^{2}MOS
    latch-mux, *see* Dual-edge-triggered storage
        element
    M-S latch, *see* Latch
CCFF, *see* Flip-flop
Circuit sizing, 106
CISC, *see* Complex instruction set computers
Clock
    buffers, 11, 108, 209, 211
    conditioning, 216
    core clock, 194, 196
    cycle, 2
    distribution, 8, 19, 119, 187, 198, 209–212
        H-tree, 24, 193, 210
        X-tree, 24, 210
    domains, 198, 210, 217
    drivers, *see* Clock buffers
    duty cycle, *see* Timing parameters
    edge degradation, 36
    energy, 180
    external, 11, 13
    frequency, 2
    gating, 112, 122, 167–177
        global, 112
        local, 113
    generation, 8, 9, 197

global clock, 193, 210, 221
grid, 24, 194, 196, 201, 209–212
hierarchy, 211
internal, 10, 12
jitter, *see* Timing parameters
load, 10
low-swing clock, 108, 177, 179
multiple phase, 8
network, *see* clock distribution
nonoverlapping clocks, 223
on-board, 10
on-chip, 10
optimal width, 69, 77
overlap, 223
phase error, 12, 197
pulsed clock, 199, 221
off-chip reference, 9, 11
on-chip reference, 179, 196
RC matched tree, 24
regenerator, 225
scan port clock, 225
single-phase, 8
skew, *see* Timing parameters
slope, 61
tree, *see* Clock distribution
tuning, 18
two-phase, 8, 70, 72
uncertainties, *see* Timing parameters
width, *see* Timing parameters
Clocking, 2
    dual-edge, 75

Clocking (*Continued*)
  edge-sensitive, 36, 64, 91
  level sensitive, 91, 98, 213
  low-swing, 108
  single-phase, 64
  soft edge-sensitive, 102
  two-phase, 70, 213
Clock-on-demand, *see* Latch
Combinational logic, 3
Complex instruction set computers, 5
Conditional capture flip-flop, *see* Flip-flop
CRAY-1, 66
Critical path, *see* Setup time violation
Critical race, *see* Hold time violation
CSE characterization, 139
Cycle stealing, *see* Time borrowing
Cycle time, 10

Data arrival analysis
  early, 65, 68, 78, 89, 93, 101
  late, 63, 66, 76, 88, 93, 98
Data look-ahead, *see* Latch
Data-to-output delay, *see* Delay
De Morgan, 25
Deep-submicron, 56
Delay, 46
  clock-to-output, 46, 85
  data-to-output, 49, 176, 179
  minimum delay restriction, 54
  insertion, 72, 95
Delay-locked loop, 12–15, 199
Design for testability, 70, 204
Deskewing, 203–205, 179–197, 199, 214
  adaptive filtering, 179
  clock spines, 192
  delay line, 192
  delay shift register, 192
  phase detection, 192, 198
DET-FF, *see* Dual-edge-triggered storage
    element
DET-LM, *see* Dual-edge-triggered storage
    element
DET-PL, *see* Dual-edge-triggered storage
    element
DETSE, *see* Dual-edge triggered storage
    element
DET-SPGFF, *see* Dual-edge-triggered storage
    element
DFT, *see* Design for testability
Digital system, 1, 8
Digital system using
  dual-edge triggered storage element, 75
  flip-flop, 63
  M–S latch, 70
  single-latch, 66

D-Latch, 26
DLL, *see* Delay-locked loop
$D - Q$ delay, *see* Data-to-output delay
DTLA-L, *see* Latch
Dual-edge-triggered storage element, 74, 113,
    179
  flip-flop, 118, 183
    symmetric pulse generator, 119, 183
  latch-mux, 116, 179, 180, 186
    C^2MOS, 181, 186
  pulsed latch, 117, 182, 186
Duty cycle, *see* Timing parameters
Dynamic hazards, 59
Dynamic logic, 216

Earl's Latch, 28
Edge sensitive, 34
EDP, *see* Energy-delay product
Effective capacitance, 106
Energy, 55, 176
  breakdown, 57
  clock energy, 165
  clocking, 58
  consumption, 55
  data and clock input, 58
  energy per transition, 166
  internal clock energy, 167
  internal non-clocked nodes, 58
  leakage, 56
  output load, 58
  short-circuit, 55
  switching, 55
Energy-delay product, 176, 185
Energy-per-transition, 58

Fanout, 62
Fast path, *see* Hold time violation
Fermi potential, 57
Finite-state machine (FSM), 3
Flip-flop, 34, 35, 159, 166, 176
  capturing latch, 35
  comparison, 164
  conditional capture flip-flop, 114, 174
  hybrid latch flip-flop, 41, 134–136, 159
  J-K flip-flop, 115
  logic equations, 39
  logic representation, 43
  modified sense amplifier flip-flop, 136–138,
    163
  pulse generator, 35, 162
  reduced clock-swing, 110
  semi-dynamic flip-flop, 123, 160, 189, 202,
    206
    conditional shut-off, 203
    dynamic, 204, 206

hold time, 202, 204
  logic embedding, 204, 205
  setup time, 203, 204
  static-one hazard, 203
sense amplifier flip-flop, 41, 161, 216
  SR latch, 216
SN7474, 37
S-R latch, 162
transparency window, 41
triggering, 37
Frequency multiplication, 10

Glitch, 43
  sensitivity, 122

Hold time, *see* Timing parameters
Hold time violation, 48, 63, 65, 76, 89
H-tree, 120, *see* Clock distribution
Hybrid latch flip-flop, *see* Flip-flop

Input transition, 59
Insertion delay, 10
Internal race immunity, 165, 176

Karnaugh map, 39
Keeper, 134

Latch 32, 180
  C^{2}MOS, 158
  clock-on-demand, 172
  data look-ahead, 169, 173
  data-transition look-ahead, 113
  hold time, 214
  logic embedding, 215, 229
  master latch, 156, 221
  M–S latch, 29, 70, 155–157, 166, 176, 184,
    190, 220, 227
    C^{2}MOS, 158
    comparison, 158, 164
    with input isolation, 157
    n-only clocked, 111, 177
  noise robustness, 167, 179, 214, 229
  noise sources, 157
  pulsed latch, 169, 172, 176, 182, 221
  setup time, 214
  slave latch, 157, 221
  split-latch, 190, 227, 229
  TSPC latch, 29, 31, 213
  TSPC M-S latch, 29, 31, 202
  with clock gating, 168, 172, 176
LC oscillator, 14
Level-sensitive, 29
Level-sensitive scan design, 29, 111, 217- 219,
  222, 229

diagnostics, 220
double latch design, 220
level sensitivity, 219
scan, 219
shift-register latch, 220
test mode, 222
testing, 220
Load
  clock, 108
  data, 108
Logic
  domino, 29, 128
  NORA, 29
  static CMOS, 128
Logic islands, 18
Logical effort, 125–127, 131, 133
  branching effort, 127
  of a domino inverter, 128
  effort delay, 126, 130, 133
  electrical effort, 126, 135
  fanout, 127, 133, 136
  multistage logic networks, 126
  optimal effort per stage, 133
  optimal number of stages, 133
  parasitic delay, 126
  pass-transistor, 127
  path effort, 127
  pull-down, 128
  pull-up, 128
  stage effort, 126, 133, 135
  of a static NAND gate, 128
  transmission-gate, 130
Loop requirement, 96
Low-swing clock, *see* Clock
LSSD, *see* Level-sensitive scan design

Machine cycle, 5
Master-slave latch, *see* latch
Micro-instruction, 5
Microprocessor
  Alpha, 25, 189, 208, 213
  Pentium, 189, 191, 193, 196
  Power4, 190, 226, 229
  PowerPC, 190, 224
  S/360 91, 28
  S/390 G4, 190, 221, 223
  UltraSPARC, 189, 200, 202, 206
M–S latch, *see* Latch
M-SAFF, *see* Flip-flop
MSL, *see* Latch
Multiplexer, 180

Nodes
  clocked, 58, 107

Nodes (*Continued*)
  dynamic, 58, 107
  nonclocked, 59, 106
  precharge/evaluate, 58, 106
Noise sources, 14
Nonoverlapping clocks, 71

ODCS, *see* On-die clock stretch/shrink
Off currents, 56
On-die clock stretch/shrink, 200
Opportunistic skew scheduling, *see* Time
  borrowing, static
Optimal setup time, 87
Optimal skew scheduling, *see* Time
  borrowing, static
Oscillator,
  crystal, 9, 18
  LC, 14
  ring, 13

Padding, *see* Delay insertion
Pass-gate, *see* Transmission-gate
Pentium, *see* Microprocessor
Perl, 138
Phase difference, 11
Phase-locked loop, 11–15, 193, 196, 201
Pipeline, 64, 67, 72, 79, 96, 100
Pipelined design, 4
PL, *see* Latch
PLL, *see* Phase-locked loop
Power4, *see* Microprocessor
PowerPC, *see* Microprocessor
Precharge/discharge, 107
Pulse generator, 35, 43, 113, 171, 184
Pulsed latch, *see* Latch

Race, *see* Hold time violation
Race immunity, 106
Race margin, *see* Internal race immunity
Race-through, *see* Hold time violation
RCSFF, *see* Flip-flop
Reconvergent fan-outs, 41
Reduced instruction set computer, 6
Reduced swing clock, *see* Low-swing clock
Resonant circuit, 9
Ring oscillator, 13
RISC, *see* Reduced instruction set computer

S/390 G4, *see* Microprocessor
Sampling window, 51
Scan test, 218
SDFF, *see* Flip-flop
Semi-dynamic flip-flop, *see* Flip-flop

Sense-amplifier, 161
Sense-amplifier flip-flop, *see* Flip-flop
Setup time, *see* Timing parameters
Setup time violation, 48, 63, 66, 76, 93
Short path, *see* Hold time violation
Signal race, *see* Hold time violation
Simulation
  setup, 130
  automated, 138
Skewed gate, 128, 135
Slack passing, *see* Time borrowing
Slow paths, *see* Setup time violation
SN7474, *see* Flip-flop
Soft clock edge, 49, 85, 189, 202
Soft error hazard, 204
Split-latch, 66
S–R latch, 28
Static inverter, 127
  FO4 delay, 138
  FO4 inverter, 130, 135
Subthreshold region, 56
Supply voltage scaling, 106
Symmetric pulse generator flip-flop, *see*
  Dual-edge-triggered storage element
Synchronous system, 3, 21

Test access port, 197
  scan, 9
TG, *see* Transmission-gate
Time borrowing, 53, 97, 210, 224
  dynamic, 91, 92
  static, 92, 95
Timing analysis
  with clock uncertainty absorption, 88
  with dynamic time borrowing, 96
  single-phase with dual-edge triggered CSE,
    75
  single-phase with flip-flop, 63
  single-phase with single latch, 66
  two-phases with M–S latch, 70
  with static time borrowing, 95
  with time borrowing and clock uncertainty,
    98
Timing parameters
  clock duty cycle, 16, 80
  clock frequency, 16
  clock jitter, 16, 83
    cycle-to-cycle, 17
    long-term, 17
  clock period, 16
  clock skew, 16, 25, 83, 191, 195, 210–212
    global, 17
    local, 17
  clock uncertainty, 83

clock uncertainty absorption, 84, 87, 103
clock width, 16, 51, 69
clock-to-output delay, *see* Delay
data-to-output delay, *see* Delay
hold time, 50, 63, 75, 89, 101, 165, 176
internal race immunity, 53
setup time, 48, 165, 176
Transistor sizing, 125, 130, 134, 136
Transition probability, 171
Transmission-gate, 155, 180, 182
Transparency window, 88, 103, 117, 183, 202
Trigger, *see* Triggering signal
Triggering signal, 34

leading-edge, 35
trailing-edge, 35

UltraSPARC, *see* Microprocessor

VCDL, *see* Voltage controlled delay-line
VCO, *see* Voltage controlled oscillator
Voltage controlled
    delay-line, 11–13
    oscillator, 11–13

X-tree, *see* Clock distribution

Printed and bound by CPI Group (UK) Ltd, Croydon, CR0 4YY

27/10/2024

14580255-0002